Praise for *Taste*

"Fascinating . . . A must for any food lover."　　—*San Francisco Chronicle*

"Mouthwatering exploration of the science of taste . . . Drawing on her experience as a professional food developer, Stuckey tantalizes readers with details about the intricacies of taste."　　—*Publishers Weekly*

"What started as a tortilla chip tasting led to an easily accessible and well-researched guide to enjoying food. Stuckey interviewed friends, chefs, and researchers and participated in empirical research by food scientists. Each chapter builds on the knowledge from the previous, culminating in some overall principles of taste and eating habits to help readers taste more thoroughly. This book will appeal to enjoyers of food, dieters, and those who wonder why the human body works the way it does in relation to food and taste."

—*Library Journal*

"A thorough investigation of the sensation of taste. Complex evaluations give readers a precise breakdown of each of these five types and how one sensation directly affects the other. Stuckey provides technical but readable discussions. . . . To aid in understanding this specialized information, the author supplies readers with many flavor-related exercises designed to increase responsiveness to various foods. A helpful, systematic approach to developing a discriminating palate."　　—*Kirkus Reviews*

"A deliciously welcome surprise. Stuckey combines accessible, well-researched information, personal anecdotes, and terrific writing to create a book that really makes the reader reconsider what that much-used word 'taste' means. It is a book that definitely made me more conscious of the way I eat and also made me feel much smarter. Both of these are good things."

—Simon Majumdar, author of *Eat My Globe* and judge on Food Channel's *Next Iron Chef*

"Simply fascinating! Compelling! A page-turner. *Taste* should be required reading for anyone who eats. In layman's terms, Barb Stuckey gives us the tools to analyze and diagnose our food neuroses, as well as get the most out of every plate of food we consume. I think if we were better tasters as a whole, we would seek out better, and yes, healthier foods for ourselves and our children. Barb confirmed for me that there is truly no need for 'kids' meals.' "

—Carla Hall, Fan Favorite on *Top Chef All-Stars*,
co-host on *The Chew*, and founder of Alchemy by Carla Hall

"A fascinating book that will change the way you think of everything you eat or drink forever."

—Kathleen Flinn, author of *The Sharper Your Knife,
the Less You Cry* and *The Kitchen Counter Cooking School*

"This book brilliantly weaves the subjective experience of eating together with the objective science of taste perception. A must-read for food lovers and cooks alike. You'll never look at a plate of food the same again."

—Ming Tsai, chef/owner of Blue Ginger and
host/executive producer of *Simply Ming*

"Understanding taste and flavor (and the difference between them) is one of the foundations of great cuisine. Barb Stuckey's book is an excellent primer on the subject. Her enthusiasm for food and science is infectious, and she explains with clarity and humor (and some neat little experiments you can try out) exactly what happens as we eat. Great reading for cooks, foodies, and indeed anyone interested in the sensory world that surrounds us."

—Heston Blumenthal, chef and owner of the Fat Duck restaurant

"*Taste* would be useful to anyone who cooks—with or without a culinary degree."

—Peter Rainsford, Ph.D., Vice President, Academic Affairs,
The Culinary Institute of America

TASTE

Surprising Stories and Science about
Why Food Tastes Good

Barb Stuckey

Previously published as
Taste What You're Missing

ATRIA PAPERBACK
New York London Toronto Sydney New Delhi

ATRIA PAPERBACK

A Division of Simon & Schuster, Inc.
1230 Avenue of the Americas
New York, NY 10020

First **ATRIA** PAPERBACK edition March 2013

Previously published as *Taste What You're Missing: The Passionate Eater's Guide to
Why Food Tastes Good*

ATRIA PAPERBACK and colophon are trademarks of Simon & Schuster, Inc.

For information about special discounts for bulk purchases,
please contact Simon & Schuster Special Sales at
1-866-506-1949 or business@simonandschuster.com.

The Simon & Schuster Speakers Bureau can bring authors to your live event.
For more information or to book an event contact the Simon & Schuster Speakers Bureau
at 1-866-248-3049 or visit our website at www.simonspeakers.com.

Designed by Carla Jayne Jones

Manufactured in the United States of America

10 9 8 7 6 5 4 3 2 1

The Library of Congress has catalogued the hardcover as follows:

Stuckey, Barb.
    Taste what you're missing : the passionate eater's guide to why food tastes good / by Barb Stuckey.
        p. cm.
    Includes bibliographical references and index.
1. Gastronomy. 2. Taste buds. 3. Food presentation. I. Title.
    TX631.S836 2012
    641.01'3—dc23
    2011038535

ISBN 978-1-4391-9073-9
ISBN 978-1-4391-9074-6 (pbk)
ISBN 978-1-4516-5472-1 (ebook)

For Roger, whom I love more than tomatoes

Contents

Introduction

A humble tortilla chip changed my life.

Most people who are obsessed with food have a different kind of epiphany. Their eye-opening, revelatory moments take place in storybook locations: a first taste of cheese made from unpasteurized milk in Aix-en-Provence. Or a fish, just plucked from the water and given a quick steam in banana leaves on a beach in Vietnam. Or a forkful of deconstructed gazpacho in Spain that made them understand—no, *really* understand—the local fascination with chilled tomato soup.

There's always a moment, but mine was much less romantic and, instead of opening up a world of flavor, it taught me just how little I knew about how to taste food.

My moment happened in a laboratory in Foster City, California, at the northern tip of Silicon Valley. In 20,000 square feet of stainless-steel lab bench tops with overhead fluorescent lighting, surrounded by homogenizers, colloid mills, dough sheeters, impingement ovens, pH meters, and tube-in-tube heat exchangers, I encountered a tortilla chip that would change my life.

I had just arrived at Mattson, the food development company where I still work as a professional food inventor. Our founder, Pete Mattson, had asked me to help with a project for a snack food company. Like many other companies, this client had enlisted our services to help them develop a new product. Our team had been tweaking the client's formula for a tortilla chip that would be sold in grocery stores. To me, there didn't seem to be much room for creativity: tortilla chips are little more than cornmeal, salt, and some kind of fat. I mean, come on. How hard could it be?

One morning as I arrived at the office one of our food technologists called me into the food lab. "Barb," she asked, "can you come taste tortilla chips?" It was 8:30 a.m.

It would take a couple of years before I'd get used to bizarre requests like this at inopportune times—a unique benefit of my job as a food developer. Could I taste frozen garlic puree at 10:00 a.m.? Could I taste meat lovers' pizza right after lunch? And would I mind a quick spot of oatmeal before heading out to happy hour?

The discussion that day was revelatory. My new colleagues debated the tortilla chip prototype, and as I listened, it seemed as if they were speaking a different language—one that I knew existed, but didn't understand. John wanted to add a touch of sugar to promote caramelization in the moisture-removal step. Teresa thought it needed a savory edge; she suggested adding autolyzed yeast extract. Pete, a self-professed saltaholic, wanted to add salt, applied topically with a bit of citric acid for zip, both ingredients ground into the finest particulate size we could achieve. *Particulate?*

The conversation turned to whether the chips should have a fresh corn flavor or a more masa harina–like flavor. The choice would determine whether or not we'd soak the kernels in calcium oxide, a processing aid that gives the corn a distinctly tortilla-like flavor that's different from the sweet flavor of corn on the cob. There was talk of using a coarser grind of milled cornmeal to affect the *mouthfeel*, a term I'd never heard. Other terms like *up-front* and *finish* were used in ways that were unfamiliar to me and I learned new ones like *rheology, mouth-melt, lubricity*, and *tannin*.

All this from three different chips fried at three different temperatures. I tasted them, but couldn't tell much of a difference between them, and so I just listened as my more experienced colleagues dissected each chip, verbalizing the nuances as if each was as distinct from each other as a slice of bread, an apple, and a chicken wing.

I wondered what I was missing. Clearly, we'd all been sampling the same chips. Why were they able to identify so many more tastes, flavors, textures, and aromas than I was? Were they just better tasters than I was? Did they have better genes? Or was it training? Practice? Experience?

This was my moment, my revelation: the tortilla chip showed me that I had no clue what was happening when I tasted food.

Later, after five or six years of working with our chefs and food technologists, I began to trust my palate and became less terrified of voicing my opinions as I tasted prototypes alongside them. I had learned the science of taste by being thrown into the frying pan of food development, shaken around for a few months, then tossed into the fire for a few more years of seasoning. Along the way, I picked up the language, a sort of Food Speak.

I even surprised myself with a newfound skill: I could take one bite of a food, consider it for a millisecond, and know exactly what it needed to taste better. For instance, I would know in less than a second if a sauce was missing acidity. More important, I knew what ingredient would give it the right type of needed sourness within the pH range we were targeting without overwhelming the other tastes and aromas.

Years later, at a client meeting, I was giving a presentation to a group of marketers at a Fortune 500 food company, trying to convince them that a combination of tomato solids and enzyme-modified cheese would deliver high levels of a taste we refer to as *umami*, making the product I was advocating irresistibly delicious. I stopped, looked at my audience, and saw a roomful of blank stares.

Umami, a taste we describe as savory, brothy, or meaty, is one of the five fundamental building blocks of flavor. Yet this group of food marketers had never even heard the term. They knew more about the tastes and aromas in wine than they knew about the tastes and aromas in food. This makes sense, though, because wine-tasting courses are common and there are hundreds of books on the fundamentals of tasting wine. Yet I'd never heard of a food-tasting course and there seemed to be no books on the subject. Why would this be? While only 34 percent of Americans drink wine, a full 100 percent of the human population eats food.

After another decade in our food lab, my taste vision got sharper and sharper. I felt as if I could see flavors more clearly, hear food more crisply, and glean more detail from everything I put into my mouth. At that first tortilla chip tasting, I had not known that there could be so many facets to a mere snack chip, yet it turns out that every food has a level of fine detail that we normally take for granted: Chips. Bananas. Tomatoes. Everything.

With my new-found sensory acuity I began to get curious about the science behind the food aromas that tempt my appetite, the tastes that hit my tongue, and the texture combinations that please my mouth more than others. I first turned to a famous resource: *The Physiology of Taste* by Jean Anthelme Brillat-Savarin, the classic tome written in 1826, often cited as the first attempt to demystify taste. I bought two different translations, but even the American English translation by M.F.K. Fisher couldn't simplify the abstract concepts enough to make them sound contemporary and fresh.

To explain how taste and smell work physiologically, Brillat-Savarin did the best he could with the science of the time, but almost two centuries have passed since he gave the world his "gastronomic meditation." The fact that he called it a *meditation* conveys pretty clearly that it's heavy on conjecture and light on science. I don't mean to blame Monsieur Brillat-Savarin—he simply had little science to reference and not much else to go on but his own experiences.

Sensory Snack

Anthelme Brillat-Savarin had no professional experience with food when he wrote *The Physiology of Taste* in 1825. He was an attorney. He proposed a sixth sense in addition to smell, taste, sight, hearing, and touch:

The last [sense] is physical love. It resides in an apparatus as complete as the mouth or the eyes . . . Although both sexes are fully equipped to feel sensation through it, they must be joined together for that purpose.

Fortunately, when I started digging into modern sensory science, I found a treasure trove of published research and institutions like the Monell Chemical Senses Center, a nonprofit organization dedicated to research on taste and smell. Called the chemical senses, taste and smell work when a chemical—in other words, a food—activates them. All food is made up of chemicals. Everything, from a freshly foraged mushroom still smelling of the earth it came from to a neon-bright Cheeze Doodle that stains your fingers orange, can be broken down into the chemical constituents that give it flavor.

chemosensory

adjective

sensitive to chemical stimuli, as in the sensory nerve endings that mediate taste and smell.

Other brilliant people all over the world, from neuroscientists and molecular biologists to dentists and psychologists, are exploring these chemical senses. I signed up to receive their professional journals and downloaded scientific papers in order to ingest the few salient points that I could understand. But as someone who had avidly avoided science classes in school, I longed to read a straightforward book written for a layperson that could teach me how to taste food without first having to teach myself science. That book didn't exist so I decided to write it.

Even before my tortilla chip epiphany, food had been the focus of my career, as well as an obsession that influenced where I vacationed, which books I read, whom I socialized with, and what I studied in school. Yet because I knew so little about taste, like most people, I didn't know how much I didn't know.

At Cornell University's Hotel School, I focused my graduate studies on food and beverage management and wine, and wrote restaurant menus for the hotels that hired us for consulting projects. Some of my culinary exposure had me cooking with Chef André Soltner, owner of the famous restaurant Lutèce in New York City. In the kind, grandfatherly Alsatian chef's class, I learned classic French techniques. With his gentle manner, he taught us how to make perfect spaetzle, using a cutting board and knife to flick them into the boiling water, how to peel a calf's brain (a skill I haven't used since), and how to cut carrots into perfect eighth-inch brunoise cubes.

But in none of these classes was I ever taught how to *taste* a carrot.

For the four years prior to graduate school I worked for Kraft in the company's food service division, which meant I was selling coffee, sauces, and other food to restaurants, and meeting chefs in their kitchens to have them taste my samples. Yet Kraft never taught me about the sensory aspects of the food I was representing. When I graduated from hotel school, I moved to San Francisco and began moonlighting as an official restaurant inspector for the *Mobil Travel Guide* in the Bay Area, eating in four- and five-star restaurants four or five nights a week. I was trained to conduct a thorough restaurant review: from judging the quality of the cocktail service at the bar to knowing whether the wineglasses were leaded crystal to checking whether the valet returned the car seat to the same spot in which the owner had left it.

Of course we were responsible for evaluating the food, but the reviewer program didn't include training on how to discern tastes or aromas or testing of my senses of taste and smell. For all my editors knew, I could have been lacking one of my senses. I wrote my reports blind to many of the sensory details of a meal.

Your Own Sensory World

Just as eyesight varies from perfect vision to nearsighted, farsighted, and grades of blindness, taste perception varies almost as widely, but we don't acknowledge the differences in the same way. There is no taste test given to children in elementary school, but all of the kids are expected to eat the same foods. Worse yet, kids are expected to eat (and enjoy) the same foods as adults. I'm not suggesting that we start to test kids' taste perception or let them eat whatever they want, but I am suggesting that we start teaching adults about the spectrum of different taste worlds that we all live in.

Taste will help you better understand what you're tasting by breaking food down to its component parts, such as the five Basic Tastes, and explaining how *flavor* differs from *taste*. Just as wine enthusiasts hone their palates with education, curious eaters—like you—can improve your tasting ability. When you understand what you taste, you will be able to better articulate not only what you like and don't like, but *why*. You'll learn how to make food taste better by understanding how flavors interact with one another and, as a result, which tastes and flavors are lacking or out of balance. This is an important skill to master if you enjoy playing restaurant critic when you eat out. And it's even more helpful if you are doing the cooking.

Cooking with a recipe is fairly simple. Dice to this size, measure this much, do this action, cook for this amount of time. But cooking doesn't necessarily explain why the recipe calls for fermented fish sauce, or how you can fix the dish if it doesn't taste right. Cooking from a recipe is really more about how to make that dish than it is about how to cook. Cooking without a recipe requires putting ingredients together using inspiration and technique, which is what they teach in culinary school. Yet it's just as important to learn how to taste, a skill you can't learn by following a formula.

In *Taste,* you will also learn tasting techniques that will help you understand what makes food delicious. You will learn to season by *taste*, not by measuring. After you understand how flavors work together, you will learn to trust your palate, freeing yourself from the tyranny of recipes. You'll understand how to season food without a guide. This can change how you feel about cooking: You could go from feeling like an overworked short-order cook for your family to feeling like an inspired artist who feeds your creations to the ones you love.

As a novice wine taster becomes more adept at identifying the flavors in wine, he tends to seek out more complex wines. The same holds true for food. I want you to seek out better food because eating better food means living a more satisfying—and arguably healthier—life. Later in the book I will explore how taste influences the food choices you make. Of course, you choose certain foods because you like them, but we'll explore *why* you like them.

Our individual food preferences change continuously from the time we are born to the time we die. Understanding this will help you better understand the food choices of your kids, partner, friends, and aging parents. The reality is that *some people are actually more sensitive tasters than others.* But those who are more sensitive tasters are not necessarily *better* tasters, chefs, or home cooks. That would be akin to saying that people with perfect vision make the best art critics. I have perfect vision, but know nothing about art. I lack the training, practice, experience, and desire to critique art. If I wanted, I could get training, practice, build my experience, and eventually develop some skill at it. But even training and being born with perfect vision wouldn't guarantee I'd be a better critic than someone with glasses or contact lenses who has a burning passion for art. Regardless of what your anatomy and genetics have endowed you with, you can be a better taster with training, practice, and a hunger to learn.

The more experience you have tasting a particular food, the better you will be able to recognize and analyze it. *Your ability to identify both tastes and smells improves with repeated exposure.* In other words, the more pinot noir wines you taste, the better you'll be able to discriminate between them: good versus bad, sweet versus dry, soft versus tannic. The same holds true for types of cheese, chocolate, apples—anything. Practice makes perfect.

For example, I used to wonder what the flavor descriptor *rancid* meant. I knew that the word referred to fats that had gone bad, but I didn't know what that smelled or tasted like. The key characteristic of rancidity is a subtle odor that is often missed. In rancid meats, it can be described as "warmed over." Rancid nuts can taste fishy. Rancid oils can smell like waxy crayons. Over the course of my tasting career at Mattson, I have smelled and tasted many rancid foods (another unique job benefit), usually in the ongoing process of tasting foods as they age (yet another benefit). In the early years I asked my more experienced colleagues to point rancidity out to me while I tasted beside them, and I learned what it was. Now when I taste spoiled fats, nuts, or meats, I know immediately that they're rancid because I've improved my ability to perceive this flavor with training and experience. *Taste* will help you enhance your perception of flavors.

As you become a better taster, you will naturally begin to pay more attention to your food. This in turn can have many benefits beyond enhancing your enjoyment of food. Gerard J. Musante, PhD, founder and director of the weight-loss facility Structure House Center for Weight Control and Lifestyle Change, says, "If you take your time while eating; if your process of consuming your meal is something you experience moment by moment; if you're truly aware of what you're doing at the table—then I believe that mindfulness will leave you more satisfied and less likely to overeat." Musante's weight-loss program focuses on teaching people how to transform their relationship with food. Through eating more mindfully, Musante says, "You begin to recognize flavors. You begin to appreciate food for what it is."

Research that links smell and taste with weight loss has already produced commercial products designed to help people lose weight. But whether you struggle with your weight or not, you can use *Taste* as a calorie-free way to get more satisfaction from the food you eat. Today, food is everywhere, but if you can feel more confident that you will derive optimal satisfaction from every bite you eat, you'll be less likely to take unmemorable bites. You won't waste precious mouthfuls on food that doesn't taste delicious to you.

Tasting is a complex process. Your preferences actually have a scientific basis and knowing this can help you understand why you eat what you eat. Or don't. Not only do you live in your own sensory world, your personal life history also affects what you choose to eat. Your food likes and dislikes are not as simple as hating *Brussels sprouts* or not loving *eggs*. If you dislike Brussels sprouts, it's possible that you are more sensitive to bitter tastes than the average taster. Or it's possible that you had a bad experience with Brussels sprouts that unconsciously (or consciously) led you to avoid them. I once met a man who couldn't drink coffee. In his childhood he'd been playing with a coffee display in the grocery store when it fell on top of him, covering him with oily, aromatic roasted coffee beans. The fear and embarrassment of this event had forever influenced the emotions he associated with coffee, leading him to avoid it for the rest of his life in order to avoid experiencing those emotions.

You use all five senses when you're exposed to food—at the grocery store, the restaurant, or the office, or when you're inundated with ubiquitous food advertising and marketing. "The senses are so influential on each other that we often don't know through which sense we're perceiving the world," says the University of California, Riverside's Lawrence Rosenblum, who studies how the senses combine and interact with one another. Once you learn what triggers each sense, you will be more aware of why you respond the way you do.

Taste will give you insider knowledge of how food marketers, restaurateurs—even farmers—leverage your instinctual reactions so you can make more informed food choices.

Mostly, I hope this book ignites a culture of taste appreciation. Somewhere along the way, we've lost touch with the sensory majesty of the meal. I'm not referring to special occasions when you dine at fine restaurants, but to the other 99 percent of your meals: the run-of-the-mill three times a day we eat at home, at work, at school, in the car. We put food in front of children and expect them to eat it, without explaining it to them, without using it to teach them a form of culinary art appreciation, and without encouraging experimentation. The best way to learn about food is to play with it!

When I was in hotel school, my mentor Tom Kelly, a professor of food and beverage management at Cornell University, encouraged frequent dining out and drinking of wine to learn more about each. It worked for me, and I believe that you, too, need to experience firsthand the concepts I write about. To help with this, *Taste* includes easy interactive exercises to illustrate the sensory concepts in the book. The exercises range from very simple (requiring only one or two ingredients) to more complex (requiring cooking). I hope you'll take the time to do them with your friends, partners, and children. Think of them as you do golf lessons, practicing piano, or taking a wine-tasting course. Even casual study can make these pursuits infinitely more enjoyable.

The Taste Revolution

Since the beginning of this century, a food revolution in the United States has been gaining momentum. We've become much more attuned to food, in almost every respect. We want to know where our food comes from. We want to know what variety of tomato or apple we're eating. We want to know the name of the farmer who grew it, as well as his farming practices and ethics. When food arrives at the restaurants where we eat, we want to know how it's handled, stored, and prepared. We want to know not only who cooked it for us, but where he went to culinary school, when he opened his first restaurant, and on what television show he first appeared.

But the revolution shouldn't stop there.

It's time that we start to understand what happens *from the plate forward*: as seen by our eyes, smelled by our noses, tasted and felt by our tongues, and heard by our ears. It's time we acknowledge that merely liking or disliking a food is biased judgment. Food appreciation is something altogether different. If we want to fully experience our food from the path it takes from our plate to our fork to the rest of our body, we need to understand the physical and psychological mechanisms of what makes up *taste*.

Yes, taste happens in your mouth, but that's only about 20 percent of the story. Food that tastes good also looks good, smells good, feels good, and sounds good. That means a lot of what we think of as *taste* comes through the four other senses. This book will explore just how intertwined the five are.

- -

sensory
adjective

1. Of or relating to the senses
2. Transmitting impulses from sense organs to nerve centers

- -

Fifteen years ago, when I was thrown into the world of food development, I wished I had a book that would teach me the most basic science behind what happens when we eat. I hope this one helps you become attuned to a world of tastes, aromas, textures, sights, and sounds—all there at every meal, free for the taking. You're about to embark upon the prerequisite text for your lifetime study of food appreciation.

Part One
The Workings of the Senses

1

Taste

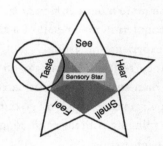

"You have bald spots on your tongue," the staff told me at a testing laboratory at the University of Florida.

At the moment of this pronouncement, my tongue was stained a brilliant royal blue. I had it smashed up against a glass microscope slide, trying to stick it out as far as it would go because I was afraid that blue saliva would run down my chin or permanently discolor my teeth. The farther I stuck it out, the less I drooled on the paper bib around my neck. I was in that ridiculous predicament so the doctoral candidate who was testing me could get a good image of my taste buds with the digital camera. As I sat in the dentist chair, I tried to hold as still as I could be expected to—with my tongue forced out, stuck to a piece of glass. Click! The enormous camera took a magnified picture and minutes later I was given the most devastating diagnosis that a professional food taster could imagine: bald spots on my tongue.

I'm going to have to make a public confession and quit my job at America's best food development firm, I thought to myself. *I will no longer be allowed to taste food professionally, which is an important part of what I do for a living. How can this be happening?*

My bald spots were clearly the result of damage, said Linda Bartoshuk, director of the Human Research Center for Smell and Taste at the University of Florida and one of the world's foremost experts on the science of taste. My heart sank further. Then Bartoshuk—spurred on by the findings inside my mouth—began to explain a bizarre taste phenomenon called the *release of inhibition.*

"What makes this particularly complicated is that another area in your mouth that doesn't have damage," she said, "may be released from inhibition and the sensations may be more intense in that area. This overshadows your bald spots. You get the counterintuitive result of a small amount of damage actually intensifying the experience of tasting."

Wait. Did I hear that correctly? The damage on my tongue might make me a more acute taster than someone without barren spots? Do damaged tongues work better than virgin ones? Should I go out and murder a few more taste buds? My mind was spinning with the conflicting input. I had no idea my tongue was going to be so interesting.

While you may be thinking that some sort of horrific accident compelled me to visit a doctor who examines tongues, the truth is, it's actually a love story.

It begins in the Sierra Nevada Mountains of northern California where one of my girlfriends and I had been skiing on the slopes of Northstar-at-Tahoe. A storm moved in and snow starting falling. Eventually a cheek-whipping wind and blinding snow sent us down the mountain to Timbercreek Inn for a fortifying glass of wine and a bite to eat. Little did we know that the storm would soon be classified as a blizzard and the roads leaving the ski resort would be shut down by the highway patrol. We were just happy that we had a coveted seat in the bar area, a glass of wine in our hands, and lunch on the way. As the storm worsened, more and more skiers accumulated in the restaurant, seeking the same refuge and sustenance. After an hour or so, we struck up a conversation with a couple of guys from San Francisco, one of them named Roger. He and I first connected over the wine I was drinking. He, too, was a fermented grape juice aficionado. We talked about our favorite grape varietals, our favorite winemakers, and our common love of the California wine country, especially Healdsburg in Sonoma County. The conversation about wine led to a discussion of our favorite restaurants in the city of San Francisco, where we both lived: Range, Ton Kiang, Myth, Delfina, Okoze, Andale, Yank Sing, and others. Six hours later we were still talking. It was time for another meal. The four of us sat down at a table in the dining room in front of a roaring fireplace. I ordered the salmon; Roger had the steak. We drank a bottle of soft, cherry-chocolate red zinfandel. We stayed at the restaurant until the highway reopened, sometime after 11:00 p.m., talking food, wine, life, and love. Our casual first date had lasted almost ten hours. We were off to a wonderful start, but I would soon learn that Roger has, ahem, *issues* with food.

An important fact about me: my education and career have been focused entirely on food. I absolutely love my job inventing new foods and making them come to life for my clients. And even though I work in the food industry, I still love to read about food and wine when I'm off the clock. I choose vacation places specifically for the food. My favorite sport is dining out. I approach cooking with enthusiasm, curiosity, and reckless abandon. When I was searching for a mate, it was important that I find a man who shared these passions. Roger seemed to fit the bill and was adorable, smart, and chivalrous. Buttercream icing on the cake.

When Roger and I went on our first official date, we ate at Café Kati, the San Francisco restaurant that in the mid-1990s made vertical food—fancy eye-popping, precariously tall plate presentations—popular. I had the roast chicken and Roger had the fillet of beef. A few days later at Tres Agaves, I ordered the pollo con mole, Roger the carne asada. Cortez: I had the lamb; Roger had the rib eye. Bistro Aix: I ordered duck confit; Roger had steak frites. Town Hall: me, sturgeon; Roger, beef cheeks. Myth: me, sea bass; Roger, beef short ribs. At some point in our courtship, I realized that this man with whom I was falling in love ate from a very limited part of the menu. To describe his menu choices, I could use the last two words that would describe my own: meat and potatoes.

When most fledgling couples have "the talk" about their future, they usually discuss their desires around starting a family, their religious beliefs, or their hopes and dreams. I had to talk to Roger about his food choices.

"How can you call yourself a foodie when all you eat is meat and potatoes?" I asked him one evening after he ordered another meat-and-potatoes entree. The other choices on the menu were so intriguing I could barely limit myself to one. People choose to live in the San Francisco Bay Area for the diversity of food choices, among other things (which are all secondary, in my food-centric opinion). Yet Roger kept choosing such bland, boring entrees that he might as well live somewhere else where the summers were sunnier, the housing prices were more affordable, and the threat of earthquakes nonexistent.

"I'm a very sensitive eater," Roger explained. "I can't handle strong tastes." Once I considered this, I realized I had seen him push aside green vegetables, flinch at having to eat salmon at a dinner party, and request that truffles—*truffles!*—be removed entirely from his plate, so that they would not contaminate his meat and potatoes. At first I attributed this to his being picky and refusing to expand his culinary repertoire. But the more I asked him to please try whatever I was eating, the more I questioned my preconceived notions about him. When he had been open to tasting my food offerings during the courtship phase, his reaction to them had often been violent. He squirmed at bitter, spicy, and sour foods. Displeasure surfaced on his face when I made him taste my vegetables. The same face emerged when he tasted a big, bold, bitter red wine. Then came my revelation one night when we were eating at Quince. I'd ordered a creative pasta dish that came with a complex sauce that I simply couldn't reverse-engineer in my head, one of my favorite things to do while I'm dining. Roger tasted it (pasta is in his limited repertoire) and proclaimed that it contained lemon zest.

"No lemon zest. Vinegar. Wine, maybe. But no lemon zest," I said.

"Let's ask the server," Roger replied. So we did. There was lemon zest in it. And from that point forward, Roger was smug in his ability to detect subtle nuances of tastes that I, the food professional, sometimes missed. As our relationship progressed, this morphed into a bit of a taste bud rivalry. Roger was good at detecting things he didn't like but he was terrible at articulating why he disliked them. He lacked the vocabulary and understanding to explain what he was experiencing in his mouth. I decided I would teach him, and that was another step down the path toward writing this book.

Bud Wiser

Fast-forward to the laboratory at the University of Florida, a steamy, sweltering summer some five years later. Roger and I, then living together, were visiting the Center for Smell and Taste because I was doing research for this book. But we also wanted to know who was the better taster. The stakes were high: bragging rights.

We entered the Smell and Taste lab by way of a waiting room, no different from the one at your dentist. After navigating a series of halls, we found Linda Bartoshuk's small group. Bartoshuk is a robust presence. Her hearty belly laugh is infectious and frequent. She's a seventy-two-year-old grandmother of five, who frequently gets so excited about her work that she has to stop to catch her breath. When I offered to treat her to dinner at any restaurant in Gainesville, she chose a mom-and-pop Asian restaurant and showed up in sensible shoes. The first time I met her at the annual Association of Chemosensory Scientists meeting, she welcomed me to the table like an old friend, even though we'd only exchanged a few e-mails. Her plate was piled high with roast beef and potatoes that she ate with gusto, talking the entire time about whom I should meet and what talks I should attend, meanwhile introducing me to her tablemates. She didn't pay much attention to what was on the plate, because Bartoshuk is less interested in what's on the plate than what's on your tongue.

By simply examining your tongue, Bartoshuk can determine your *taster type*. With a glance, she'll know whether you're likely to be a Supertaster, a term she has coined to describe people like Roger. She dyes the surface of your tongue with blue food coloring, then looks for the taste buds, which don't absorb the blue dye as much, and as a result show up as turquoise bubbles against a vivid royal blue background. Bartoshuk and her colleague Jennifer Stamps examine your tongue for the density of those turquoise taste bud bubbles distributed across its surface.

This is the surface of University of Pennsylvania professor Paul Rozin's tongue. He is anatomically a Supertaster due to the density of taste buds, which show up as bubbles against the dark blue-stained background. Jennifer Stamps calls his "a beautiful tongue."

The poster child of tongues belongs to Paul Rozin, a professor of psychology at the University of Pennsylvania. Almost the entire surface of his tongue (shown here) is covered with turquoise taste buds. Anatomically, he is a superlative Supertaster. If you're curious about your own tongue and don't have access to a smell and taste center (some are listed at the back of the book), you can count the number of taste buds in a certain area of your tongue, specifically an area the size of a notebook paper reinforcement label. These doughnut-shaped stickers are used to repair tears in the antiquated form of media known as paper. This taste bud exercise is easy to do at home with blue food coloring and a reinforcement, detailed for you in at the end of the chapter. Stand in front of a mirror over a sink. This is important, lest you spot-stain your rug with blue food coloring as Roger and I did in our home. Using a cotton swab, dab the blue dye onto your tongue until it's good and blue. Try to keep it sticking out or your lips, too, will be dyed. Place the reinforcement on your tongue and count the number of round bumps that show up inside the little hole.

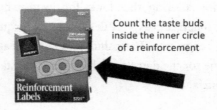

Count the taste buds inside the inner circle of a reinforcement

If you have up to fifteen taste buds inside the inner circle of the reinforcement, you are probably one of Bartoshuk's Tasters or even a Nontaster. According to Bartoshuk, if you have forty buds or more, you are likely to be a Supertaster. The quantity of taste buds on your tongue has been correlated with the intensity at which you taste most things. In other words, the more buds, the more you taste. Other research has shown that Supertasters also experience other things as more intense. Salt is saltier, sweet is sweeter. Bitter for you is bitter*er*.

"Supertasters are one end of a distribution," says Bartoshuk, meaning that there's a typical bell curve distribution of tasters in the population. People at the left end of the range have limitations to their ability to taste certain things. Bartoshuk called these people Nontasters: 25 to 30 percent of the population falls there. Supertasters reside at the right end of the distribution, comprising another 25 to 30 percent of the population. Supertasters may taste the same food three times more strongly than a Nontaster would. There's a huge range of abilities that fall in the middle and these people—about half of the population—Bartoshuk called Tasters.

I like Bartoshuk immensely, but I dislike the term Supertaster as strongly. It conjures up the image of a superior breed of being, dressed in blue tights with a cape, leaping over tall buildings. It promises super powers and super

The Distribution of Taster Types

The area under the curve contains the entire human population.
Tolerant Tasters make up about 25 percent of the population.
Tasters make up the majority, and HyperTasters make up about 25 percent.

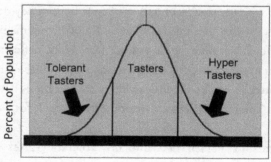

Density or number of taste buds

insight. It also assigns judgment: something that's super must be better than something that's not. Being a Supertaster doesn't make you better at anything (such as cooking or appreciating wine) except tasting. It's no more "super" than being born with perfect vision. As for the term Nontasters, it is not only insulting, it is misleading. The people Bartoshuk classifies as Nontasters *can* taste most things; it is mainly bitter things that they cannot taste. I prefer to use different terminology to describe the difference in human taste perception.

Taster Types

I call the groupings of tasters *HyperTasters* (in lieu of Supertasters), *Tasters*, and *Tolerant Tasters* (in lieu of Nontasters). HyperTasters are on the far right end of Bartoshuk's bell curve. These people are very sensitive to the tastes on their tongue. A small amount of something will tickle their taste buds powerfully. They usually have very strong likes and dislikes because they taste things so intensely; food is often overwhelming, in both good and bad ways. They are often focused, maniacally, on food. Others, ostensibly those at the sensitive end of the HyperTaster grouping, may eat a bland diet, having been burned too many times by the strong bitter or sour tastes they've experienced. Roger says this is why he eats mostly "safe" foods.

The opposite of HyperTasters are Tolerant Tasters, who are at the other end of the spectrum. These Tolerant Tasters may barely notice the taste of a food at all because it's so mild to them. They're usually tolerant of a broad range of flavors and foods. Tolerant Tasters are at the far left end of the distribution curve with the smallest number of taste buds. They don't sense a lot of strong tastes, so they don't usually have a lot of strong dislikes. They may drink their coffee black because they don't taste black coffee as bitter. They may choose intense, bitter red wines because to them, these wines don't taste so intensely bitter. They're much less likely to be obsessed with food than people like me. Tolerant Tasters are the most fun to cook for—they complain the least.

Linda Bartoshuk herself falls into my Tolerant Taster grouping. She describes herself as "an extremely insensitive nontaster," as I probably would have guessed after eating two meals with her, had she not told me. She's picky, but this is driven almost entirely by her food allergies.

If anyone ever tries to tell you that you don't have a palate that's discriminating enough to know much about taste, you can always tell them that one of the (if not *the*) world's leading taste experts is a Tolerant Taster. Case in point: you don't have to be a HyperTaster to be an expert on taste.

Tasters make up the largest percentage of all tasters. They fall in the center of Bartoshuk's distribution. Keep in mind that, although the majority of the population (40 to 50 percent) falls into this category, Tasters can run the gamut. People who are Tasters can have almost no taste buds on their tongue, meaning that they don't experience much intensity (these Tasters border on being Tolerant Tasters). But other Tasters can have almost as many taste buds as HyperTasters, so they, too, can be excessively sensitive. It just depends where Tasters fall within their grouping: in other words, where they fall within the largest, middle group of tasters.

I have met married couples or pairs of siblings, one of whom would eat just about anything while the other one had a laundry list of things he or she wouldn't eat. We might assume that the ones with the limited diets are not very "good" at tasting or that they are less open-minded than people with a larger repertoire of food. Both of these assumptions are far from universally true and may in some cases be false. Often, people who have very limited diets are HyperTasters. While this may seen counterintuitive, it makes perfect sense. To illustrate, let me use a different sensory discrepancy between Roger and me.

My hearing is much better than Roger's. While he attributes this to having attended too many Dire Straits and Tom Petty concerts in his youth, I know that his father suffers from the same subtle loss, so it's most likely genetic. Sometimes I'll walk in on Roger watching television with the sound so loud that I worry the neighbors are going to complain. He and I experience the same level of decibels quite differently. I have to remind myself of this when I'm cooking for Roger. To him, the bitter vegetal notes of butternut squash are excruciatingly loud. To me, they give this sweet, starchy vegetable a pleasant complexity. We've learned to compromise. He turns the sound down while watching the tube and I no longer make him eat green vegetables that aren't drowned in hollandaise to temper the bitterness he experiences from them.

Props to PROP

When the idea of segmenting people into groups based on their taster type was still new, Bartoshuk discovered that she could distinguish between the HyperTaster type and others by measuring people's ability to taste a single bitter chemical called PROP (6-n-propylthiouracil), pronounced "prope." To HyperTasters, this chemical tastes horribly bitter. To Tolerants, it has no flavor at all. What a fantastic discovery this was at the time! This meant it was possible to simply ask people to taste a glass of water with a few drops of PROP in it and, if they reacted violently, you could anoint them HyperTasters. If they had a moderate reaction to it, you would call them Tasters, and if they tasted nothing, you'd call them Tolerant Tasters. This was a relief because it was so much easier than asking people to taste samples of hundreds of different foods and standardizing their responses, as researchers had to do to determine taster types in the past.

As is often the case in science, later work demonstrated that it wasn't quite so simple: There are HyperTasters who cannot taste PROP, and some Tasters who can. It turns out that individual differences in taste sensation are much more complicated than initially thought. There are three main factors that account for individual differences in ability to taste. The first factor is the anatomy of your tongue, which is measured by counting your taste buds. The second factor is your medical history, and the third factor is your genes.

Extractions, Infections, and Accidents, Oh My!

In addition to the density of taste buds on your tongue, the second indicator of taster type is your medical history. There are many things that can happen to you that can affect your ability to taste.

For example, an ear infection can damage the chorda tympani taste nerve, which runs from the tongue up through the middle ear to the brain. Viruses, including flus and herpes, can also damage this nerve, resulting in the death of innocent taste buds and taste bud bald spots. This is a great reason to get a flu shot every year and to treat earaches immediately (not to mention that treatment relieves the often-excruciating pain). If you had serious ear infections as a child, it's likely that creamy, fatty, and fried foods send you over the moon because, if your chorda tympani taste nerve was damaged, your trigeminal nerve (the one that carries texture information) may be now be singing loudly without inhibition. I am somewhat certain that Roger's childhood ear problems gifted him with a love of foie gras, burgers, cheese, and ice cream.

Accidents, especially head injuries, can also result in loss of some taste (and smell) function. For this reason, if you value your sense of taste, always buckle your seat belt; wear a helmet when you bike, ski, or skate; and forgo head-banging sports like American football and boxing.

Wisdom tooth extraction, a common surgery, occurs precariously close to the chorda tympani taste nerve. If wisdom tooth extraction surgery goes wrong, it can damage taste irreparably, even though that may not be the dentist's fault.

Another perpetrator of crime against taste buds is disease. Parkinson's disease, for example, may result in the loss of the sense of smell. And decreased ability to smell is one of the harbingers of Alzheimer's disease. If you have any of these conditions, you may not be able to taste certain things even if you have a heavy concentration of taste buds on your tongue. In other words, being an anatomical HyperTaster doesn't mean you can taste more than Tasters or Tolerant Tasters. It's not so black and white.

Surgery and dental work might explain the bald spots on my tongue. When I was a child, I had a benign cyst on the underside of my tongue. It was removed during outpatient surgery and I rarely think much about it, except when it's about to rain. One of the strange results of this surgery is that I can detect changes in barometric pressure with my tongue. When the clouds are about to burst, my tongue starts to throb—my own, internal barometric pressure gauge.

One of the other results of my surgery, Bartoshuk and her colleagues think, was damage to the taste buds on my tongue. The location and existence of my bald spots seem to indicate damage to my trigeminal nerve, the nerve that carries pain information from my mouth to my brain. The phenomenon I mentioned earlier, the release of inhibition, means that when this nerve was injured, its ties to the taste buds in a certain area of my tongue were clipped. These abandoned buds eventually withered and faded away, leaving behind bald spots. This allowed other areas on my tongue the freedom to communicate taste information a little bit louder, without inhibition.

Roger's tongue showed some damage too, probably due to oral surgery he had to remove his uvula to widen his airway. His uvula surgery did two things that directly benefit me. First, it relieved his most bed-shaking snoring. Second, it resulted in the loss of some of his taste buds, thereby giving him a bit of humility when it comes to our taste rivalry.

It turns out that Roger and I are both HyperTasters, as proclaimed by Bartoshuk's graduate student Jennifer Stamps after eight hours of thorough evaluation of our tongues. Was she sure about me, I asked, given my bald spots?

She responded, "Your intensity rating to the PROP test paper was definitely a HyperTaster rating: a 90. I know you have trigeminal damage because when the nerve endings degenerate, they take out the taste buds they once surrounded and leave behind holes and bare spots. Your ratings for taste on the tip of your tongue may be lower than before your loss of taste buds but they were decent for the ones you have left, which indicates your chorda tympani taste nerve is working fine."

Whew. This was a great, huge, relief. *Thankfully*, I thought to myself, *I have absolutely nothing to hide with regard to my professional fitness.* Roger and I flew back to California together, radiating relief from having been anointed HyperTasters.

It wasn't until about a week later, after a glass of wine or four, that Roger again brought up the trip to Florida. Perhaps I deserved it, having doubted his tasting ability in some capacity. Or perhaps it was just his competitive side flaring up again. Regardless, he mischievously hinted at a truth he was concealing for my own benefit.

"Are you sure you want to know?" he asked me of this mysterious fact he possessed.

"Of course I do," I said, not at all sure at this point if I did.

"You must know that I'm only going to tell you this for your own good. I mean, since you're writing this book about taste and all."

Then he said, "When you went to the bathroom while I was in the taste tasting lab with Jennifer, she told me something about you."

My heart stopped beating. I held my breath.

"She told me I have more taste buds than you," he said, looking somewhat sheepish, somewhat swaggering. The worst part about it was I knew he was right. Household taste bragging rights belong to Mr. Meat and Potatoes.

Blame Your Parents

The third factor for taster type is genetics, which is also responsible for the traits you get from Mom and Dad. You may have the genes that allow you to taste something such as the bitter chemical PROP. Or you may not. It's simple genetics at work. For each trait, such as blue eyes or the ability to taste PROP, you get one gene from your mother and one from your father. If PROP tastes extremely bitter to you, it's likely that both your PROP-tasting genes (one from Mom and one from Dad) are turned on. If you can't taste it at all, it's likely that both your PROP-tasting genes are turned off. And if you have a mild reaction to PROP, you may have one gene that's on (the one from Mom or the one from Dad) and one that's off (the other one, from the other parent). In other words, you need at least one "PROP on" gene to react to the compound at all, but two copies of the gene in order to have a HyperTaster reaction to PROP. Keep in mind, however, that your reaction to PROP is not a universal indication of your reaction to other compounds. Just because you can't taste PROP doesn't mean you can't taste PTC (phenylthiocarbamide, another bitter chemical used to test for HyperTaster type), the bitterness in Brussels sprouts, or the bitterness in beer. In fact, "There are PROP nontasters who are Supertasters [HyperTasters]," says Bartoshuk, just to confound the matter.

Your unique tongue anatomy and genetics combine in an interesting way that results in your own unique experience of food. If you have the genetic ability to taste PROP but a low density of taste buds, you may taste food very differently from someone with a high density of taste buds who cannot taste PROP. And how these traits affect your food choices is even more complicated when you layer on the other senses. HyperTasting has no correlation whatsoever with hypersmelling, a fact that doesn't get enough attention.

Measuring Taste

Because of all this complexity, Bartoshuk defaults to a more straightforward test for identifying a HyperTaster: she simply asks how intense some things taste. When we were in Florida, we sampled popcorn, lasagna, peanut butter, and grape jelly. But herein lies the rub. How in the world do you measure this? What would happen if you gave both Roger and me a plate of roasted butternut squash (with sage and brown butter, please) and asked us to rate it? How would you know that his perception of bitterness is the same as my perception of bitterness? Or sweetness? Or saltiness?

The fact is that perception occurs in the mind. As a result, it is virtually impossible to measure accurately. Take the perception of beauty. How pretty is Angelina Jolie? How beautiful is the city of Paris? There is no definitive answer to either of these questions. The answer varies with the individual. Asking about someone's perception of food is similarly complicated. Each person is biased (and informed) by his own anatomy, genetics, and life experience. Taste perception is in the mouth and brain of the beholder.

The solution, says Bartoshuk, is to peg taste intensity to something that we *can* measure, such as sound. She conducted an experiment that started with a can of Coca-Cola, a product with a standardized recipe (or formula, as we say in the food development business) for the United States market. Because of this, a Coke in your neighborhood, city, or state will taste exactly the same as a Coke in mine (though it is different in Mexico, where Coke is sweetened with sugar instead of high fructose corn syrup). This wouldn't be the case if she had used tomatoes or berries or beef, which vary by location, variety, season, and how they've been stored. Bartoshuk asked consumers to rate the sweetness of Coke on a scale in which the bottom of the scale is "no sweetness at all" and the top of the scale is "the sweetest you've ever tasted." In this first test, almost everyone put the sweetness at the same place: about two-thirds of the way up.

"You look at that and you think, *Wow! People's senses of taste are really very similar,*" says Bartoshuk. But this is where things get interesting. In a second phase of the test, she outfitted the testees with earphones and a sound dial. In this round of testing, Bartoshuk used a technique called *cross-modality matching.* Here, a *modality* is a sense. Crossing modalities means using one sense (hearing) to gauge another (taste). She asked the tasters to adjust the volume of the sound they heard through their earphones to match the sweetness of the Coke. The results showed that the group of HyperTasters adjusted the sound up to the level of a train whistle, or about 90 decibels. Conversely, the group of Tolerant Tasters adjusted the sound level down to that of a telephone dial tone, about 80 decibels. A difference of 10 decibels equates to a factor of two. In other words, "That tells us through this matching technique that people with the most taste buds experience twice the sweetness," says Bartoshuk. Maximum sweet for a HyperTaster is twice as intense as maximum sweet for a Tolerant Taster.

"PROP remains a very good way to identify taster types; if you taste PROP and you are one of the individuals who get a very strong bitter taste, you know that you are an anatomical Supertaster. However, some of those who cannot taste PROP can also have the anatomy of Supertasters; they just don't taste PROP." This is all to say that there's no litmus test for determining your taster type. It's confusing, to say the least.

What's also confusing is that the food choices of HyperTasters and Tolerant Tasters are highly unpredictable. If at this point you're starting to wonder if you are doomed to a life of blandness because you may or may not be a HyperTaster, fear not. Being a HyperTaster is as much a curse as it is a blessing. Think of all the amazing bitter foods Roger and others like him simply can't eat. In times of famine or shortage, Tolerant Tasters would be able to sustain themselves on bitter roots and plants while Roger would wither away and die without his meat and potatoes, unable to tolerate the bitter greens he'd be forced to subsist on.

You may be a HyperTaster and not recognize Roger's behavior as your own. In fact, Paul Rozin, owner of the "beautiful tongue" depicted earlier, makes food choices very different from Roger's. He, too, finds many bitter foods almost unbearable, but he has come to like many of them, mostly through repeated exposure and coming to appreciate the strong sensation they give him. He considers himself an omnivore, and has traveled the world seeking out unique food experiences. I, too, am a HyperTaster and almost an omnivore. I seek out bitter foods like the aforementioned Brussels sprouts. Bitter foods taste bitter to me, but I love the sensation. Roger avoids it. I have to be superattuned to flavor nuances for my job and I'm pretty good at detecting these subtle flavors, lemon zest possibly excepted.

I asked Bartoshuk how scientists are able to make correlations between types of tasters and the food choices we make. Trying to answer this question, Bartoshuk seemed to share my exasperation as she gave reasons why two people of the same taster type might choose very different diets. For example, I may have put a lot more time and thought than Roger into the concept of Brussels sprouts, eventually developing an appreciation of them. People who are brought up in cultures that believe bitter foods are good for them generally end up liking bitter foods. Conversely, if someone has a terrible experience with Brussels sprouts, such as vomiting right after eating them, that person may tend to avoid them because that experience conditioned him to do so, consciously or unconsciously.

"The truth is, you just can't make predictions for one person," said Bartoshuk. "It's just too complicated. We can statistically do a very good job. You give me a hundred subjects in one group and a hundred in another and I will be able to tell you some things about their *average* behavior that will be right on. But I will miss by a mile with individuals."

This is the ultimate problem with these groupings of taster type. After people learn about taster types, they seem to want their type to explain why they eat what they eat. But we are complex creatures, each of us living in our own individual sensory world, each of which is colored by a combination of anatomy, medical history, genetics, culture, and life experience. The best way to describe the type of taster I am is that I am a Barb Taster. And Roger is a Roger Taster. That makes you a [insert your name here] Taster.

The bottom line is that your taster type is just one factor among many in why you make the food choices you do.

You can't change the anatomy of your tongue, just as you can't change your eye color or height. But a height limitation doesn't mean that you can't teach yourself to be an excellent basketball player. And everyone—including you—can teach himself to be an excellent taster.

Smell, See, Hear, Touch

Everything you've just read is the tip of the iceberg of what we casually refer to as *taste*.

Taste, taste buds, and the tongue represent a tiny amount of what you experience when you eat food. A smidgeon. An itsy bit. Not a whole heck of a lot.

This is because of the fact that your tongue can taste only a few things, namely sweet, sour, bitter, salt, and savory. There's absolutely no way to prove how much information the tongue contributes.

Perhaps 10 percent is attributable to taste and 90 percent to smell, but only if you're dividing the entire experience of eating between just those two senses of taste and smell. What about the other three? When you add the influence of touch, hearing, and sight, things get really interesting. Our experience with food—which we simply call *taste*—is actually a multisensory adventure.

First, I'm going to teach you how the senses work. From there we'll explore each of the Basic Tastes, the nuances of flavor, and finally, how everything comes together. Deliciously.

A Visit to Taste & Smell Central

I was in Philadelphia in a minivan heading to the restaurant Buddakan with five researchers from Monell, a nonprofit research institution focused on uncovering the scientific mysteries of taste and smell. Seated behind me was Marci

Pelchat, whose expertise includes food cravings and food addiction. One of the chattier scientists from Monell, she pointed out landmarks during our quick ride. Reading Terminal Market, Pelchat told me, houses downtown Philly's version of a farmers' market, although it has become a tourist destination.

"But I think you can still get pickled tongue there," she said.

"Beef tongue?" I asked, remembering it from the Jewish delis of my youth, where the sight of a five-pound cow's tongue would make me squeal.

"Kosher tongue?" inquired Bob Margolskee, a Monellian who studies taste at the molecular level.

"Yes, you know, cow tongue that's been cured like corned beef. You slice it to make sandwiches," said Pelchat. "I bought a whole one there to use as a prop for a demonstration on taste that I was giving a while ago. It's an amazing way to point out the papillae on the tongue. Just like ours, only bigger."

"I'm in need of a tongue," chimed in Michael Tordoff, a researcher at Monell who studies, among other things, our taste for the mineral calcium.

"You mean a *human* tongue?" I asked. "For research?"

"Yes," he answered. Human tissue samples are apparently hard to obtain.

"I'm not just looking for any tongue," said Tordoff, "I'm looking for a *fresh* tongue."

"Some people at Monell just lop off their own taste buds," Margolskee told me as we arrived at the restaurant. When you dine with sensory scientists, disturbing visual images about their work accompany the meal.

The first thing I learned when I got to Monell was how the improper use of the word *taste* sends sensory scientists into a bit of a tizzy. I was corrected no fewer than five times for using the word *taste* to mean the combination of taste, smell, and texture. Science demands proper terminology, but since I'm not a scientist, I don't use their jargon. I speak and write in plain English, as you do, and I say things such as *I can't taste anything when I have a cold and my nose is stuffed up.* Yet my taste system, as the scientists pointed out, is most likely perfectly functional. It's my sense of smell that is compromised when I have the flu. Most of what we think of as taste is smell. Some of the food odors we smell come from sniffing the food when it's under our nose (outside the mouth). But most of the aromas we perceive when we eat are released in the mouth and reach the nose through the mouth.

When you eat something new, you taste it for the first time, although you'll also smell, feel, and touch it. When someone asks you whether you like a food, he asks if you like the taste of it, but what he really wants to know is if you like its combination of smell, taste, texture, appearance, and sound. Yet *taste* has become the default word for the experience of eating food—in both noun and verb form—because we do (using correct scientific terminology) taste with our mouth.

You instinctively know that what you experience when you eat is just as dependent on your nose as on your tongue. In fact, research has proved that every other sense—sight, hearing, touch, and smell—can influence what you taste as well. But you don't eat with your nose. You don't put food into your ears or eyes. When the system is working the way it should, you put food into your mouth.

This causes us to connect flavor to the mouth because it's the place where we taste, the place where taste sensations are initially sparked. But only a small portion of what you experience as flavor happens geographically on the tongue. Linking the entire experience of food to the mouth, though understandable, is what causes the confusion.

There are only five tastes that humans can detect using their mouths alone. Technically speaking, if it's not one of the five Basic Tastes, it's not a taste at all. Everything else we experience in the mouth is either an aroma or a texture. The combination of these three characteristics—tastes, aromas, and texture—is correctly called *flavor*. The tastes in a tomato include sweet, sour, and umami (the taste described as savory or brothy). The aromas in a tomato include grassy, green, fruity, musty, and earthy. The texture depends on how ripe the fruit is and how it has been prepared, from juicy, firm, raw tomatoes to tender, soft, simmered ones. And the overall flavor of a tomato is what you know of as a tomato, the whole gestalt.

To appreciate, firsthand, how profound the difference between taste and smell is, I suggest you try the exercise called Separating Taste from Smell, which is at the end of this chapter. Plug your nose, and while holding it shut, put a jelly bean in your mouth and start chewing. After a few chews, you'll easily detect the two Basic Tastes evident in it: sweet and sour. Once you release your nostrils, the aromas of it will spring forth: tropical, cherry, pear, melon, buttered popcorn. The flavor of the jelly bean you've chosen is the combination of the two Basic Tastes, the signature aromas of the variety of flavor you've chosen, and the texture, chewy-tender.

Of course, you don't have to use a jelly bean to isolate the taste from the aroma of a food. Use a cherry tomato or fig or strawberry and you'll experience the same thing. With your nose pinched shut, you'll detect very little of the characteristic flavor of what's in your mouth. You'll get only sweet, sour, bitter, salt, or umami. Release your nostrils, breathe, and then you will get the aromas of tomato, fig, or strawberry.

In *Taste*, I'm going to use plain English and say, "when you taste a tomato" even though I may be talking about the total multisensorial experience of eating a tomato. But I will also use (and recommend the common usage of) the term *savor* as a verb when the word *taste* is scientifically incorrect. For example, "When you savor a tomato, you get the green aroma first, followed by the basic tastes sweet and sour." We usually think of savoring something as consuming it with delight. But Merriam-Webster defines the verb *savor* as "to have experience of," so it's perfect in sentences where the word *taste* is incorrect.

The linguistic tendency to use the word *taste* to mean flavor is not an idiosyncrasy of the English language. University of Pennsylvania professor Paul Rozin asked bilingual speakers of nine languages to provide synonyms for the words *taste* and *flavor*. They were given a dictionary to see if they could find better words. And then they were educated on the difference between the Basic Tastes and aroma. In seven of the nine languages (Spanish, German, Czech, Hebrew, Hindi, Tamil, Mandarin Chinese), it appears that this same idiosyncrasy exists, so that if it goes in the mouth, it's *tasted*. Only Hungarian and French seemed to have words that hinted at a distinction between the concept of taste versus that of taste plus aroma: what you know now is flavor.

The word for flavor in French is, not coincidentally, *saveur*.

Sensory Snack

Taste and smell are the only two senses we confuse. Imagine someone saying, "When I heard that Renoir, I was really moved." or "I like to watch the radio." It just doesn't happen.

The Five Basic Tastes

Once you learn the five building blocks of taste, you will see how they work in harmony with the other senses and start thinking more critically about what you're tasting. Four are familiar to most people: sweet, sour, salt, and bitter. The fifth, umami (pronounced ōō-mä'mē, which rhymes with "who MAH me"), is a newer term, imported from Japan, which is loosely translated as savory, brothy, meaty, delicious, or round. Umami refers to the savory taste of certain amino acids that make a good beef steak or soup stock taste so rich and full. If you were to take all the salt out of chicken or beef broth, you'd be left with umami. It isn't a taste we crave on its own. It really needs to be paired with salt and aromas. More about this complicated taste later in the book.

Think about the five Basic Tastes as the five tips of a star. Throughout the book I'll be using the star as a tool to help you form a visual representation of how inextricably linked each taste is with the others, as well as how important all five senses are when you're experiencing food.

The Taste Star: The Five Basic Tastes

The Sensory Star: The Five Senses

I use the star shape because it's perfectly balanced, which is how I think about the five tastes: there isn't one taste that's more important than the others for making food taste good. But not every food should contain all five Basic Tastes. And not every food should contain all five in equal proportions. Take wine, for example. Most wines contain the sour and bitter tastes. Some wines are sweet. But almost no wines are salty. And this is a good thing: it doesn't belong.

When you're cooking or seasoning a dish, it is important to make sure that one taste doesn't dominate the others, whether all five tastes are present or not. When one taste (or aroma) dominates, we say that the dish is out of balance; the way the star would be if one of the points were bigger than the others. A wine that tasted salty would definitely be out of balance.

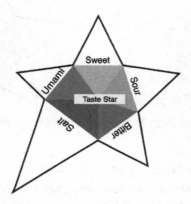

When one taste is out of balance, it throws off the whole food, dish, or drink.

It's fairly easy to recognize a dish that's out of balance from too much salt or bitterness, because it will be unpleasant (as a salty wine would be). What's harder to identify is a dish with too much umami or savoriness. When you become more familiar with umami, you'll be able to tell when there's too much of it. Let's review the five Basic Tastes very broadly; then for each Basic Taste we will go into more depth in its own chapter.

Sweet

Sweet is the term we use for simple carbohydrate compounds such as sucrose, more commonly known as sugar. Almost universally, people describe sweet tastes as pleasant. While sugar is the purest form of this taste, lots of other things naturally taste sweet, such as fruit (which contains fructose) and dairy products (which contain lactose). Sugar is a quick source of calories, so we are genetically predisposed to seek out sweet things.

Sour

We use the term *sour* to describe the taste of acidity. Lemon juice and vinegar are two of the most prevalent sources of sourness in food; both liquids are high in acid (citric acid in the case of lemon juice, acetic acid in the case of vinegar). Acidity is usually pleasant but can quickly become unpleasant at high levels; a squeeze of lemon can brighten up the flavor of grilled fish or a glass of iced tea, but straight lemon juice is mouth-puckeringly unpleasant. Some people, however, love the extreme sourness of lemons so much that they suck on lemons repeatedly. This can cause the enamel on their teeth to erode if they do it often enough for a long enough period of time. In a pretty nifty design—compliments of Mother Nature—most people find that foods with tooth-rotting acid levels are too sour to eat.

Some acids make foods and beverages taste fresh and bright, whereas other acids indicate spoilage and can trigger instant rejection of those foods. Acids also help preserve some foods, such as pickles.

Bitter

Bitter individual tolerance varies more widely for bitter foods than for any of the other Basic Tastes. Bitter foods can be very unpleasant on their own if they are not balanced by other tastes and flavors. Coffee, tea, and red wine are common bitter beverages that can be delicious when carefully crafted. Most compounds with medicinal effects have a bitter taste—some at low levels, some at high levels. Our ability to taste bitterness has evolved to help us identify substances that can be toxic. Caffeine, for instance, is extremely bitter. It has a very real, well-recognized medicinal benefit—stimulation—but at high levels it can be toxic. Many poisons taste bitter and their medicinal effect—death—is one you probably want to avoid. That's why it makes sense that humans have a complicated, distrustful view of bitter tastes.

Salt

Salt is the term we use to describe the taste of sodium ions. The most common form of salt is sodium chloride, which we add to food while cooking or sprinkle on at the table. Many foods naturally contain sodium, such as seafood and celery. Salt is critical to life, but we cannot store excess sodium in our bodies, so we are programmed to seek it out in the form of food. In modern times, getting just enough sodium in our diets—without excess—has proved to be a bigger challenge than getting too little. Regardless of how much sodium we consume, our craving for salt is natural—and critical to survival.

Umami

Umami is the most difficult taste to explain because the term is not commonly used outside the world of food or outside Japan, where the term originated. Umami is the taste of glutamates—amino acids that are present in some foods such as beef and mushrooms. The best-known umami-rich compound is glutamic acid—or glutamate—which occurs naturally in some foods such as mushrooms and seaweed. Monosodium glutamate (MSG) is the salt of glutamic acid and this form is often added to foods as a seasoning. We sometimes describe umami as tasting meaty, savory, satisfying, or full. Think of the difference between raw ground beef—which has little umami—and a well-cooked hamburger, which has lots. Other savory foods that are high in umami are cooked tomatoes and the king of umami: aged Parmesan cheese.

The Geography of the Tongue

How do we actually taste these five Basic Tastes? One possibility is that different regions of the tongue process different tastes—as on this map, some version of which you might have seen in elementary school.

The taste map of the tongue. Be careful how you interpret this!

The map shows the geography of the tongue and which area corresponds to which taste. People love this anatomical map because it makes some sense of the multitude of things you taste simultaneously in your mouth when you eat. There's a major problem with it, though: it's completely misleading. It seems to say that you can taste only one of the five Basic Tastes on one area of your tongue. *This is not true.* You can taste all five of the Basic Tastes on all parts of your tongue. Certain tastes will be more intense in certain areas, but that doesn't mean you can't detect these tastes elsewhere. Sour is really intense on the side of the tongue but you can taste sour everywhere. Prove it to yourself now by doing the Sour All Over Tasting exercise (at the end of the chapter): dip a cotton swab into distilled vinegar, a really tart liquid. Dab the swab around your mouth without swallowing. You should taste sour all over your mouth, not just on the sides of your tongue. That is, unless you have bald spots on your tongue or other damage to your taste nerves.

Breaking Down the Sense of Taste

Once you put food in your mouth, there are four dimensions of the sense of taste, according to Paul Breslin of Rutgers University and the Monell Chemical Senses Center. He calls the five Basic Tastes *qualities*. I like to think of the five Basic Tastes as the first question (Q) of taste, the What: What is it you're tasting? Sweet? Sour? Bitter? The taste qualities of a tomato are sweet and sour.

The second taste dimension is intensity, or the degree of magnitude of a taste. I think of this as the How: How intense is the taste? How strong? How weak? Examples of this would be an extremely sweet tomato and a mildly sour one.

The third dimension he calls *oral location*, or the Where: Where in the mouth or throat is the taste perceived? Again, the best example of oral location is that most people detect sourness most strongly on the sides of the tongue. As you just learned, you can taste each of the five Basic Tastes everywhere. Where you perceive them to be the strongest is relevant but probably won't affect your enjoyment of food.

The final dimension is the timing, or the When: When do you sense the taste? When does it start? When does it end? When is it the most intense? You may describe the timing of the tastes of homegrown cherry tomatoes as bitter and green at the beginning, as you bite through the outer skin. This would be called the *initial* or *up-front* taste. Then, as you keep chewing, you may experience sourness next. That would be the *middle* of the taste experience. And last, as you continue to chew and swallow, you may experience the sweetness. This would be the taste in the *finish*. The timing of taste is wonderfully illustrated by how differently we perceive the sweetness of sugar versus that of artificial sweeteners. Even though artificial sweeteners taste sweet, each one—sucralose, aspartame, saccharin, stevia, and so on—is detected in your mouth either faster or slower than sugar, and each lasts for a different length of time after the sweetness of sugar would have cleared from your mouth.

To really understand the way sweetness works for you, do the Sweetness Profile tasting exercise in the chapter on Sweet. I'll explore this in more detail there.

The Four Qs of Taste: What, How, Where, and When

What?	How?	Where?	When?
Type	Intensity	Location	Timing
Basic Tastes	Magnitude of the taste sensation	Perception of where in the mouth/throat the sensation occurs	When the taste is sensed
Examples:	Examples:	Example:	Example:
Sweet, sour, bitter, salt, umami	Mildly sweet or extremely sour	Sourness perceived more strongly on the side of the tongue	Sourness at the beginning, with a lingering bitter taste in the finish

Adapted from: Paul Breslin, *Human Taste: Peripheral Anatomy, Taste Transduction, and Coding*

Tasting food with your mouth is called *gustation*. This word comes from Latin and shares its origin with the word *gusto*. I love the simple redundancy of the term *eating with gusto*, which means "gustation with gusto"—a good phrase to help you remember the scientific term for taste. Smelling aromas is referred to as *olfaction*.

$$\text{Gustation} + \text{Olfaction} + \text{Texture} = \text{Flavor}$$

We'll get into more later about each of these building-block tastes, as well as flavors, but for now, let's talk about how your mouth works.

Born-Again Buds

In 1999, when I was fairly new to my job at Mattson, I had a client in the vegetable business. The owners hired us to come up with exciting new vegetable appetizer ideas for their restaurant customers. After a few days of thinking about the assignment, I knew at least one of the ideas I wanted to create: cornmeal-crusted fried green tomatoes like the ones my father had cooked for us every summer weekend of my childhood. Typically, you make fried green tomatoes with unripe, green fruit that are harder and less juicy than ripe red ones. But because they're usually sliced before they're fried, the tomato slices are wet and flimsy and would be too difficult for our client to handle in the quantity needed for restaurants. As I worked through the idea in my head, it morphed and emerged as cherry tomatoes—much easier to handle. When we couldn't find green cherry tomatoes, I decided to start experimenting with red, ripe ones just to see how they might work out. We call this the *proof of concept phase.*

Marianne Paloncy, one of our best chefs, called me into the food lab to show me samples of the inaugural batch of my creation. This is my favorite part of my job: seeing and tasting the physical manifestation of an idea. The little cherry tomatoes dipped in batter and lightly coated in cornmeal were adorably cute and promised crunch and flavor. They'd make a perfect restaurant appetizer.

Paloncy dropped a handful of them into the fryer basket and we waited two minutes for the cornmeal to crisp up on the outside, while the little spheres bobbed around in the bubbling oil. When she pulled them out of the fryer, they were glowingly golden brown. I couldn't resist. As I reached my hand into the fryer basket, Paloncy started to speak, but before I could register her warning, I'd already popped a tomato and was pressing it against the roof of my mouth with my tongue. The 375°F frying oil, which had heated the copious water inside the red (ripe and juicy!) tomato, exploded in my mouth with excruciating force and volume. I opened up instinctively and spit out the entire thing, along with a huge flap of skin that I'd burned off the roof of my mouth. I could barely talk. *There goes my tasting career,* I thought.

Luckily, my damaged palate healed and the thousands of cells I'd scorched off my tongue and the roof of my mouth were replaced within two weeks, the normal amount of time it takes taste cells to regenerate. In fact, cells are constantly turning over from normal wear and tear. This programmed cell death makes perfect sense, since our taste buds are built to be abused, says Breslin: "If you could make a car that could regenerate parts for you, the thing you'd want to regenerate would be the treads on your tires." You could say the lava-hot tomato burned the rubber off the roof of my mouth.

My accident gave me an appreciation for the mouth's resilience and new insight into how important it is to have a mouth that functions properly. Much of the pleasure in eating comes from taste and texture, two things that were compromised (or painful) while I was healing. Still, I was shocked by how quickly my sense of taste came back. The mouth is one of the most important tools we have to ensure our survival as a species. At the most basic level, if we don't eat, we can't nourish ourselves. And if we eat dangerous things, we can poison ourselves. If our sense of taste were to fail, we'd be at great risk.

Adaptation

We experience the bitter Basic Taste most strongly at the back of the tongue. If swish-and-spit tasters doesn't allow the food to fully saturate those bitter-sensitive buds, their perception will be misinformed. Nonswallowers might argue that in certain circumstances, it may not be possible to swallow everything you taste. In some wine competitions, judges have to taste 100 or more wines in a single day, and one can only imagine the state they would be in if they'd swallowed even 100 small sips of each wine. My counterargument would be that these competitions make their judges taste too many wines in a single day. No one, no matter how good a taster she is, can discriminate between that many wines. A phenomenon called *taste adaptation* sets in after your tongue has been exposed to too many tastes in a short interval.

With each additional sample of a taste, you become more and more adapted to that taste: this means that you require more and more of it to get a similar level of intensity. Michael O'Mahony of the food science department at the University of California, Davis, writes,

> A constant odor or taste stimulus will be perceived as decreasing in intensity while sensitivity to that stimulus is also decreased. For sensory evaluation, this poses problems. It means that a taste or odor has a tendency to vanish while it is being observed and that sensitivity to subsequent stimuli will be altered. Such sensitivity drift in the human instrument must be anticipated in the design of measurement procedures for the sensory evaluation of food.

And wine, I argue. The phenomenon of taste and smell adaptation gives overly intense wines an advantage if they are tasted late in the day, but puts them at a disadvantage if they are tasted early. Vice versa for subtle wines. If only the "human instrument"—our mouth, tongue, nose, eyes, ears, and brain—were less prone to the failures and foibles of the human condition.

Adaptation is also the reason you can't taste your own saliva. Your saliva contains sodium and potassium chloride, which make it slightly salty. Yet I'm sure you don't think of your mouth as having a salty taste. That's because you're adapted to it. Your taste cells are in constant contact with it twenty-four hours a day. In fact, the makeup of your saliva is changing all the time, but in such small increments that you don't notice it. It takes a rapid increase in the concentration of salt in your mouth to wake up your adapted taste buds so that you recognize it as salty. This happens when you eat.

Your own saliva is possibly the only thing in the world that you will perceive as having zero flavor. Even water differs in composition from your saliva, so you experience water as having some sort of taste. You've probably even said something along the lines of "It tastes like water."

Chef Grant Achatz, of Alinea restaurant in Chicago, serves a twenty-three-course menu of small bites. This addresses the impact of adaptation, which he refers to as the law of diminishing returns. He understands that people don't need twenty-four ounces of steak to satisfy their craving. When talking about his tasting menu he says,

> That's why the steak is only two ounces. By your fifth bite, you're really done with that steak. You know what it's going to taste like. The actual flavor starts to deaden on the palate. If we were to make you take ten more bites, by the time you got to bite fifteen, the steak's just not that compelling anymore. So if we have a series of twenty-three small courses, where it's a burst of flavor on the palate, then you move on to something completely different . . . and then completely different. That helps us set up a more exciting meal.

Conscious, Unconscious, and Conscientious Tasting

When something tastes wrong, your body usually won't let you swallow. That's a good thing: it means your taste system is effectively serving its role as gatekeeper of the body. When and if you swallow, your sinuses get a burst of flavorful vapors from the wad of food your tongue forces to the back of your throat. Technically the wad is called a *bolus*, although that term somehow manages to be even more unappetizing than wad. Again, this will happen only if you're breathing, something that I recommend you do constantly while you chew, and carefully when you swallow. You will continue to taste the food as long as the volatile aromas are being drawn back up into your nose through the normal course of chewing, breathing, and swallowing.

And that's it. The conscious tasting part of eating is over. To summarize, you derive the initial pleasure of tasting food only while it's in your mouth and throat. This is pretty obvious when you stop to think about it. But that's the problem: we don't often stop to think about savoring food. We're too busy reaching for the next mouthful. If you really want to taste something, it's a good idea to keep it in your mouth as long as possible. Put your fork down. Take a few breaths. Chew some more. Swish it around. Then swallow. From that point forward, your food follows a very well-documented, studied path through your digestive tract.

I called the previous process the *conscious* part of tasting because scientists have recently discovered that we have taste cells much farther along in the digestive tract. There are cells in your stomach, small intestine, and pancreas that look and act exactly like those in your mouth. This was a surprise at first, says Monell scientist Bob Margolskee, but it makes sense: the gut needs to be able to identify food in order to know what to do with it.

"Our stomach and our intestines want to know what we've consumed. Then they can respond in the appropriate way by turning up the digestive juices," he says. This phenomenon is known as *gastrointestinal chemosensation*. Even though we are not really conscious of tasting food much farther than our throat, our gut "tastes" nutrients and responds accordingly.

Sensory Snack

Some things can potentiate others, or mask others. For example, the sodium laurel sulphate in toothpaste makes orange juice taste very bitter. The sodium laurel sulphate masks the sweetness, which then potentiates the sourness and bitterness in the juice.

All of the sensory processes I've described thus far have one thing in common: their connection to the brain. When it comes to taste, this connection is an everyday matter of life or death. Making the wrong everyday choices with other senses, such as what music to listen to or what image to look at, may harm you, but probably won't kill you. But making the wrong choice about what to eat can be lethal. Your taste system is set up to give important information to the brain instantaneously so that the body can react accordingly. To quote Monell scientist Danielle Reed, "Tasting is deciding."

Of course, the system has its flaws. Many poisonous mushrooms are reported to be delicious, although the person who reported this probably ended up experiencing liver or kidney failure, a common side effect of the toxins that can be fatal. Or a rogue mushroom eater might develop a conditioned aversion to them in the future (if he survived). With a conditioned aversion, an eater associates a particular food with a bad outcome, and then avoids the food consciously or unconsciously. A not-so-careful wild mushroom forager might love the taste of those death cap mushrooms as they're going down the first time, but he may develop an aversion to—a dislike of—all mushrooms in general as a result of vomiting or otherwise getting sick from them. He'll probably be most averse to the straw mushroom, which looks like the death cap's identical twin. It's amazing what puking—or liver or kidney failure—will do to your food preferences. An old adage captures this survival mechanism: there are old mushroom foragers, and bold mushroom foragers, but there are no old, bold mushroom foragers.

I call conditioned aversions The Tequila Effect. Anyone who has drunk too much tequila and subsequently prayed to the porcelain god knows of what I write. For a period of time after you get sick from a certain food, you will recoil at the mere smell of it. In fact, it often takes disguising the offensive food to get you to ingest it again, such as by adding lime juice and triple sec. Salting the rim of the glass helps too.

Another way Mother Nature can fool us is with dangerous compounds like botulinum toxin, which are flavorless. While lower doses administered topically under the brand name Botox can give you an unnaturally smooth forehead, higher doses ingested orally can result in muscle weakness, paralysis, and even death. These exceptions to the rules of taste are rare, however. Most wholesome, safe food tastes good. Most spoiled or poisonous food tastes bad. Your sense of taste gives you information as it gives you pleasure.

The Taste Committee

I have seen lots of depictions of the taste system, technical illustrations of the tongue and taste buds that are tough to decipher. But you probably don't care about discerning the difference between the fungiform papillae (the taste buds at the front of your tongue) and the vallate papillae (those at the back). To describe the physiology of taste without getting too bogged down in the science, Monell's Danielle Reed gave me one of the best analogies for how taste works: it's like a committee meeting in your mouth.

Imagine your mouth as a meeting room full of business colleagues who serve on the taste committee. Each has been elected to represent others like themselves in the organization: a true democracy. These colleagues work together to tackle projects (food). The committee members get together frequently to discuss new projects (incoming food) and report to the boss (your brain). The five people on the committee represent each group of the taste team: Sweet, Sour, Bitter, Salt, and Umami. One or two of the members usually dominate the meetings. Sometimes they all get a word in. It depends on the project (the food). The hard work of the committee doesn't really come together until a member of the Retronasal Olfaction Team sweeps through the meeting room, like the flow of aromas from your mouth to your olfactory receptors. This is when the work gels.

Unless we pay really close attention to what we're eating, or unless one of the tastes or flavors is out of balance, we usually don't think about each taste and each aroma separately. We react to them all as if we've been sent a summary report from our taste and smell systems. *Pepperoni pizza* is our first conscious reaction to taking a bite of a favorite food, not *sweet, salt, sour, umami tastes combined with tomato, Parmesan, cured meat, and green herbaceous aromas* . . . ah, *pepperoni pizza*. We gloss over the details, but our brain fuses the information into one coherent packet of information.

We know a lot about reactions to things like pizza and other compound foods. We don't know a lot about what happens at the level of the taste bud.

Each taste bud contains many taste receptor cells, and some of these receptors detect bitter, some detect sweet, and some detect umami. Taste cells are specialized to detect only that one taste. Sweet, bitter, and umami tastes are detected when they latch onto a taste receptor in a hand-in-glove fashion. Scientists have yet to identify the ones that detect salt. Sour and salt have to pass through ion channels to be detected. It's a harder system to study, which is why we have yet to identify the salt receptor. The fact that both are detected similarly is one of the reasons people confuse salt and sour, as they do with really sour or really salty foods.

The taste cells in the mouth and throat are connected to three main nerves in the head called *cranial nerves*, which carry taste messages to the brain. Taste cells on the front of the tongue connect to a cranial nerve called the chorda tympani. There's a similar body part in the ear, the tympanum, both of which share a Latin root with the tympani drum. The chorda tympani, carrying taste information to the brain, passes right through the middle ear. The glossopharyngeal nerve connects the back of the tongue.

The taste receptor cells that function as the "rubber tires" of your mouth relay information to nerves that tell the brain about what you are tasting. Even though they're around for less than two weeks due to programmed cell death "tread replacement," the new taste cells that replace them reestablish a connection to the brain. The fact that this connection is cut and reconnected—continually—over a course of ten days adds credence to just how important the taste system is.

There are other sensations that we perceive as tastes that aren't really tastes; nor are they smells. In fact they are related to the sense of touch—the same sense that detects pain. These sensations connect to the trigeminal nerve, which also transmits touch, pain, and the indication of temperature (what you feel when you drink something hot). Other examples of trigeminal sensations in the mouth are the cooling of mint and the spicy-hot of chile peppers. When you eat a salsa that's spicy-hot, you might say it tastes hot, but that would be incorrect (now I'm thinking like a scientist). You actually feel spicy-hot chiles. The trigeminal nerve is also the main facial nerve involved in classic migraine headaches. Not surprisingly, due to their shared relay to the brain, spicy foods are a common trigger of migraines.

We don't know where the signals go once they move along the relay system from the taste receptor cell to the nerve to the brain. Perhaps one day we'll have a cool map of the brain that marks sweet with a star, the way a map of the United States marks Washington, D.C., as the capital. But for now, we don't know a lot about how taste works in the brain. The boss keeps his secrets well hidden.

Taste: Separating Taste from Smell

I'm asking you to do this experiment with your eyes closed. Read it thoroughly up to the spoiler alert before attempting to do it. That way, you'll know what to expect.

YOU WILL NEED

A bowl of jelly beans of various flavors (I am partial to Jelly Belly brand because they have very complex, realistic flavors)

If you are opposed to eating candy, I suggest you use a basket of mixed bite-size fruit such as grapes, strawberries, raspberries, and blueberries.

DIRECTIONS

1. Close your eyes and pinch your nose shut with one hand, so that you cannot breathe through your nose.
2. Without releasing your nose, put your other hand in the bowl, mix up the contents, pick one piece, and put it in your mouth without looking at it. (The idea is to remain unaware of what you are putting in your mouth.)
3. Begin to chew, slowly, without releasing your nose. Keep chewing.
4. Keep chewing, and without releasing your nose, think about what you taste. Sweet? Sour? Can you tell what flavor it is?
5. Release your nose.

OBSERVE

1. When your nose is pinched, it's likely you'll taste only sweet and sour. That's because those are the only two Basic Tastes that exist in most jelly beans and fruit. With your nose shut, your tongue—or gustatory system—is doing the best it can by identifying which Basic Tastes are present.

2. What happens when you release your nose is that the volatile aromas from the jelly bean escape back up through your nasal passages, into your olfactory system, where you experience smell. This system has a lot more than just five smells to work with. There are reportedly thousands of aromas that humans can detect.

Taste: Your Taster Type

YOU WILL NEED

Blue food coloring (Caution: The blue dye stains fabrics, including carpets and towels!)

Small nonporous cup

Paper towels

Cotton swabs

1 paper or plastic reinforcement ring for each person

Magnifying glass

Mirror

DIRECTIONS

1. Pour a bit of blue food coloring into a nonporous cup.

2. Using a paper towel, blot your tongue to remove as much saliva as possible.

3. Dip the swab into the food coloring and apply the dye to your

tongue. Let it saturate your tongue and dry out before you apply the reinforcement ring. Try to keep your tongue out while you're doing this, or the ring will get wet and messy!

4. Apply a reinforcement ring to your blue tongue.
5. Using the magnifying glass and mirror, count the number of round taste buds inside the inner circle.

RESULTS*

    0–15 = Tolerant Taster

    16–39 = Taster

    40 or more = HyperTaster

* REMEMBER! Tongue anatomy is only one of the determinants of your taster type. And remember, this test measures only your sense of taste. It doesn't take into account how good your senses of smell or touch are, which you'll learn team up with taste to create the concept of flavor.

Taste: Sour All Over

YOU WILL NEED

    1 cup distilled white vinegar (any kind of vinegar will do, but
        distilled has the cleanest one-dimensional sour taste)

    Shallow bowl or ramekin

    1 cotton swab for each person who will taste

    Saltine crackers and water for each taster

    A handheld mirror for everyone who will be tasting

DIRECTIONS

1. Pour the vinegar into a shallow bowl or ramekin.
2. Dip the cotton swab into the vinegar.
3. Swab your tongue/mouth in distinct places, in order, being

careful not to swallow or close your mouth until you've swabbed and tasted the spot you're exploring. You may want to break this up into four exercises.

    a. Cleanse your palate with a saltine and water. Swab the middle of your tongue.

    b. Cleanse your palate with a saltine and water. Swab the sides of your tongue.

    c. Cleanse your palate with a saltine and water. Swab the back of your tongue.

    d. Cleanse your palate with a saltine and water. Swab the insides of your cheeks.

OBSERVE

1. As you touch the swab around your mouth, notice what you experience.
2. What you are experiencing is the sour taste. Most people experience the sour taste on all tissues in the mouth.
3. You've just disproved the taste map diagram!

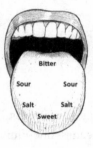

It's easy to prove that you can detect all tastes on all parts of the tongue. What's harder to prove is that this theory is completely wrong, which it is not. There *are* slight differences in the intensity at which we taste things in each area of the tongue. This truth is where the taste map originated.

2

Smell

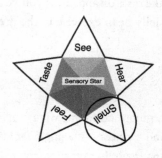

Volatiles are where the action is.

Harry Klee, University of Florida

If you stop people on the street and ask them which one of the five senses they'd be willing to give up, the most likely response will be smell. In fact, *The Escapist* magazine conducted an online poll of its readers and the results from 772 respondents were unequivocal: smell would be the first to go. But people don't understand that what we taste is largely what we smell. While the taste committee's role is critically important, it is smell that gives us pleasure from foods. After all, your flavor world without smell would contain only five things.

Like all high-performing organizations, our smell and taste "teams" need to work closely together to accomplish their shared objective: recognizing what's in our mouth and making a decision about it. Consider just how closely taste and smell work together: retronasal olfaction is all about smelling through your mouth. You simply cannot dissociate the two. Which sense is doing the heavy lifting? Which deserves the credit? Which has the power?

Scientists estimate that between 75 and 95 percent of what we "taste" is actually smell. So that smell committee member is responsible for a lot more of the final product (the brain's perception of taste) than the taste committee. Smell is so important to the business of taste that losing your sense of smell can practically make taste disappear. If you've ever had a cold, you know this. Or, if you did the jelly bean exercise in the previous chapter, you experienced it.

Sensory Snack

Winemaker Ilja Gort, renowned for his wine "tasting" skills, took out an insurance policy on his nose—making headlines and hammering home the fact that the majority of tasting comes from the sense of smell.

The Two Ways We Smell

When you put food in your mouth, your sharp *Homo sapiens* teeth come together to tear it into smaller pieces. Chewing, also called *mastication*, is what we do to prepare food for digestion. Chewing increases the surface area of the food, so that the enzymes in our body can start to release the nutrition from it. If you were to swallow food whole, you'd eventually digest it, but your digestive system would have to work much harder. One of the reasons we have such sharp teeth is to jump-start the process of getting energy from food. Another benefit of pulverizing the food into tiny bits is that we get to enjoy the flavor of it while we chew. The enjoyment of food reinforces our behavior and we eat again, which insures that we get proper nutrition.

Taste and the Two Ways We Smell

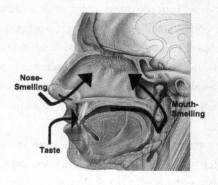

When you crush food between your teeth, tongue, cheeks, and the roof of your mouth (the soft palate), the aromas are released and get sucked up through your nose as you breathe. This flow of aromas from your mouth to your nose is called *retronasal olfaction*. This term is a mouthful (pun intended) so I'm going to refer to it as *mouth-smelling*. It's different from the type of smelling you do with your nose when you sniff something that's outside your body. Technically, smelling through your nose while the food is still outside your mouth is called *orthonasal olfaction*, but I'm going to refer to it as *nose-smelling*.

One important thing to remember about these terms is that in both cases, your nose is where the smell processing occurs. There are no olfactory receptors in your mouth. But in mouth-smelling (retronasal olfaction), your mouth is where the aroma molecules came from, on their way to being processed in your nose. Please don't ever say, "I smell with my mouth." This statement is untrue. You smell with your nose from inside your mouth by means of retronasal olfaction.

Sensory Snack

The human jaw is the only joint in the body where the left side and the right side are unable to function independently.

When you see people slurping wine, they're doing it to increase retronasal olfaction, or mouth-smelling. Slurping in air while tasting increases the flow of aromas, allowing you to smell and taste more—more quickly. While chewing with the mouth open is a social taboo in the United States, doing so would actually help us savor better because it would increase mouth-smelling. I will beam like a proud teacher if this book results in an eating public that slurps more unabashedly when eating and drinking. The more slurping you do, the more flavor you get.

You start to taste your food when the compounds in it change form in your mouth. With a crunchy food like a potato chip, you don't start to taste until it mixes with your saliva and starts to break down. Saliva moistens dry food and helps release tastes and aromas. There are enzymes in saliva that break large molecules into smaller ones that have more flavor. The very makeup of saliva helps you taste.

The second way you taste is when a soft food like chocolate starts to dissolve from the heat of your mouth. One of the most seductive qualities of good chocolate is that it melts precisely at human body temperature, which provides a textural experience unlike any other food. This fact makes chocolate one of nature's most perfect foods.

Adding moisture or heat—or both—to any food helps liberate volatile aromas from the food so you can experience mouth-smelling, which is where you experience most of the aromas from food. For instance, neither the potato chip nor the chocolate bar has much aroma on its own in its room-temperature state outside your mouth, even if you get really close and inhale deeply. It's not until you put the food in your mouth that the moisture and heat make it become the crave-able food you know it to be.

The quantity and quality of saliva that you produce enhance your sense of taste. The autoimmune disease Sjögren's syndrome causes the moisture-producing glands to shut down and patients often report that they experience a loss of taste. This is because they have a dry mouth, trouble chewing, and difficulty swallowing, all of which interfere with important contributors to the flavor of food.

How does the information from the aromas and tastes that are released from food make it to your brain? Not surprisingly, the process begins on your tongue, the upper surface of which is covered with taste buds. How many buds you have determines your taster type (HyperTaster, Taster, or Tolerant Taster). Each taste bud contains millions of cells, most of whose surfaces are covered with taste receptors—proteins that recognize molecules in food and communicate that information to the cell itself, which in turn sends signals through nerves to the brain. While most receptor-bearing cells are in the taste buds, they are also present on the roof of your mouth, on the sides of your mouth, and in your throat.

There are horrific accounts of people who have had their tongues cut out who sometimes report still being able to taste, although swallowing is a problem. Swallowing is critically important to tasting because it triggers mouth-smelling. I'm skeptical of people who say they can savor a food or wine fully by swishing it around in their mouth and then spitting it out. In my professional opinion, you miss nuances of flavor when you don't swallow. Mouth-smelling continues as you swallow food, so spitting it out cuts off the flavor process perception from its natural progression.

Taste by Numbers

Your tongue draws the outline of the food with the five Basic Tastes, the way a paint-by-numbers drawing looks before you fill in the sections. You can glean only so much detail from your tongue, though, because its palette is made up of only five tastes: sweet, sour, bitter, salt, and umami. Eating a meal without your sense of smell would be like viewing a great painting as an early, unfinished sketch. Without smell, you would experience coffee as bitter water. Milk would be slightly sweet water, and lemonade would be simply sweet and sour.

Because you cannot smell a food unless its aroma volatilizes or evaporates into the air, smell molecules are called *volatiles*. Some foods, such as citrus fruit, contain lots of volatiles. Others, such as salt, have few or none. Simply applying heat can help release the volatile aromas of food: A loaf of bread on the counter doesn't have much of an aroma—but put it in the oven for a few minutes and soon your kitchen will smell of fresh-baked bread as the loaf's volatile aromas are released by the heat of the oven. This is a great trick for getting more aroma out of just about anything. If it's lacking in smell, put it in the oven for a short period of time to release some volatiles. I do this with all bread, regardless of how fresh it is. Unequivocally, warm bread is better than the alternative because it has more volatiles and hence more flavor (not to mention other yum-producing texture and flavor changes). It never fails to amaze me how few restaurants take advantage of this simple act, which can increase sensory input for practically zero extra money or effort. I also put crackers, tortilla chips, and potato chips in the oven sometimes, just to shock the aroma molecules awake. You have to be careful doing this, though, as they're dry to begin with, so they burn easily. I heat them for about two minutes in a preheated 400°F oven.

Once the potent aromas of the food you're eating travel from your mouth up to your nose by means of mouth-smelling, you'll experience the signature flavor of the food. It's the aromas of a food that fill in the lush, saturated detail of the work of art.

At RN74, a bustling San Francisco restaurant, I recently enjoyed a perfectly cooked snow-white halibut in a beurre blanc sauce. Without my sense of smell, I would have experienced only two Basic Tastes in the sauce: sour from lemon juice and wine plus salt from salt. When I recognized the aromas of the citrus, the oaky, buttery fermentation aroma from the chardonnay, and the dairy notes from the butter that combine to make this classic recipe, it became a fully realized beurre blanc sauce.

The estimate that 75 to 95 percent of what we taste is actually smell lines up pretty well with the illustration below. The black outlines represent a very small percentage of the volume of the picture. The real beauty comes from the

The Five Basic Tastes Draw the Outlines   Aromas Fill in the Details

combination of the Basic Taste outlines from your tongue and the majority of the picture, which is what they give structure to: smell, from both your nose and mouth.

When I swallowed a bite of the halibut with the white wine and butter sauce, I also swallowed the smell of it. Literally. The aromas that you detect in your nose are actually physical things. Many people think smell is like sound, which we cannot see or touch, but this is not at all true. You can capture sound electronically or digitally, but you can't capture sound particles. In contrast, you can capture the "particles" of smell—the volatile molecules that enter the air surrounding an object—even though they are submicroscopically small. You pull them into your body with each breath. If you smell it, you ingest it. This is a wonderful image to keep in mind as you consider an exquisite plate of food in front of you, such as the halibut with the beurre blanc wafting odors of the ocean and butter and wine into your body. It's a bit more disturbing when you consider the smells you encounter in a bathroom or behind the exhaust pipe of an eighteen-wheeler. Have no fear, though, as smell molecules are generally not absorbed into the bloodstream.

Your Nose, the Tool

The inside of your nose is lined with a very thin layer of mucus. It's always there, even if it isn't dripping out of your nostrils or down your throat from a cold or allergic reaction. This mucus layer is full of tiny hairs, or cilia, that wave back and forth in the mucus in a motion similar to the way human hair or seaweed moves when it is underwater. Once a smell is captured in the mucus, these hairs wave back and forth and flush the smell particles down your throat. If you're breathing (which I will continue to advocate throughout the book), you'll further enhance your ability to detect the flavor of the food by smelling it not only through your nose orthonasally, but also through your mouth, retronasally. The cilia also act as antennae that help hold onto the aroma particles as they are absorbed into the olfactory receptors inside the nose. All of this happens almost instantaneously, resulting in the recognition of a smell: *That's a tomato.*

When you have a cold, this always-present mucus layer gets thicker. If your cold is bad enough, the mucus may be too thick to allow smells to penetrate it. If a smell can't penetrate the mucus layer, you won't detect it. If you want to savor more while you're sick, you could take a decongestant, which would thin and dry the mucus so that smells are able to penetrate it once again. But as long as one of your nostrils works, you can smell. Even though you have two of them, research has proved that you can't tell which side of your nose is doing the smelling. So don't sweat it if only one side is stuffed up. Your brain won't know the difference.

Once a smell penetrates the mucus and binds with a smell receptor cell at the top of the nose, there's an electrochemical change in the cell that fires off a message to the brain. The smell nerve is cranial nerve I, which runs up through the bridge of the nose directly into the brain cavity.

Nerves in the Head That Play a Role in Taste and Smell

Cranial Nerve	Name	Information Carried
I	Olfactory	Smell
V	Trigeminal	Texture, heat, and pain
VII	Chorda tympani (branch of facial)	Taste from the front of the tongue
IX	Glossopharyngeal	Taste from the back of the tongue

The discovery of olfactory receptors was so hungrily anticipated and important that the two people who found them, Linda Buck and Richard Axel, won the Nobel Prize. Here, Axel compares the complexity of smell to sight:

> In the eye, we can discriminate several hundred different hues, and we do so with receptor molecules that recognize different wavelengths of light but we only have three such receptor molecules. In the olfactory system, there are at least a thousand . . . There are a thousand [olfactory] points in the brain. A given odor will activate a "set"—what we call a combination—of those points, so that the quality of an odor would be determined by a spatial pattern of neuroactivation in the brain. Every odor we have examined has a different signature that is represented by the spatial pattern of neurofiring in the brain.

This "odor signature" also communicates with the brain region where your memory of odors is matched with the odor you're smelling. The many different experts on olfaction I've talked to generally agree that our preferences for smells are learned. Clearly, we can learn to associate a smell with a certain experience.

Fragrant Flashbacks

I grew up in Baltimore, Maryland. The Inner Harbor of the Chesapeake Bay, which is Baltimore's downtown hub, used to be home to McCormick, the world's largest spice company. The McCormick spice processing facilities operated downtown while I was a child, releasing into the city aromatic, spiced air, which hung over downtown like a wonderful cloud of kitchen aromas. It wasn't until I was much older—at work—that I realized that my hometown had a distinct smell. One of our employees at Mattson was busily emptying out old containers of spices and had dumped a bunch of random spices into a trash can. Imagine a mixture of cinnamon, thyme, black pepper, cumin, oregano, dill, celery salt, cardamom, clove, and a dozen other dry spices. I happened to pass the trash can and when I got a whiff of the contents, I froze. With one inhalation, I was transported home, to my childhood, standing in the Inner Harbor with my parents. Since I'd left Baltimore, I hadn't encountered that odd, completely unintentional blend of spices.

This is the power of aroma: the ability to make a grown woman stick her head into a trash can and tear up. Smell can move us in this way because of our anatomy: olfaction is the only sense that does not first pass through the brain's sensory switchboard, the thalamus. It essentially shortcuts its way straight to the boss man without having to deal with his minions. A very powerful sense, indeed.

The first time a signal from a smell reaches your brain, it etches a signature into your memory. I experienced what I call a *Fragrant Flashback* standing over that trash can of spices. A wonderful scene in Disney's *Ratatouille* also illustrates this concept when the most wonderfully named food critic, Anton Ego, digs into a dish of chef Remi's piping hot vegetable ratatouille and is instantly transported back to his childhood. This is not because childhood is where we encode our smell memories, but because Ego was a child when he first experienced ratatouille and I was a child when I first smelled the aroma of dozens of spices mingled together. If you smell ratatouille for the first time at age thirty-seven, you will connect it to that thirty-seven-year-old period of your life the next time you smell it. Smell memory is so strong it can take you back to the first time you experienced a food. It just so happens that we experience most food for the first time when we're young.

Personal history, culture, and learning determine our smell preferences. Linda Bartoshuk says that our sense of smell acts as an emotional sponge:

> If you sniff an odor and some predator takes a bite out of you, you're going to learn that predator is bad news. Even better, you eat something with an odor, now if you get calories from the food that odor was in: wow! That food is registered in your brain as something very good for you. Something bad happens to you, the olfactory signal is disgusting to you; something good happens to you, you love that smell.

These responses are called *conditioned preferences* or *conditioned aversions*. You are conditioned to like or dislike a food in part by your experiences that accompany ingestion of it. It's also similar to the response Pavlov was able to condition in his dogs. The dogs associated the sound of a bell with being fed. Eventually the bell, alone, could make them salivate.

But even if you and your blood sibling grew up in the same ratatouille-eating household, this doesn't mean that you'll both have the same reaction to the dish's aroma. In addition to our personal histories, our sense of smell is as much a matter of genes (or nature) as it is how we're brought up (or nurtured). Our genes determine how intensely we can smell things in the same way that they make some of us HyperTasters and some Tolerant Tasters.

Some people are genetically predisposed to specific anosmias: the inability to smell a specific compound, or smell-blindness. A few of these are widely known—for example, after eating asparagus, many people notice that their urine has a strong odor that is somewhat vegetal, sulfurous, petroleum-like, and tinny. Others are completely oblivious to this smell. Almost everyone who eats asparagus produces urine with an "off" odor, but if you're lucky, you're smell-blind to it. Specific anosmias exist for many other odorous compounds as well.

Nose-Smelling Versus Mouth-Smelling

Now head of a lab at Monell Chemical Senses Center, Johan Lundström was born and raised in Sweden. He was always intrigued with smells, but it was his pet German shepherd, Ella, who was responsible for his professional path toward olfaction. On walks, Lundström noticed that Ella would detect other dogs' urine in places where he could not smell a thing. He knew that dogs have a more acute sense of smell than humans, but he was particularly interested in her behavior after she'd smelled another dog's output. Sometimes her tail would go wild; sometimes she'd get anxious; at other times she'd be amorous.

"She definitely got some kind of signal from that urine," Lundström says. Watching his dog react to the secret signals in other dogs' urine got him interested in pheromones, the aromas that emanate from many living creatures, reportedly used as signals. Lundström eventually went to graduate school to earn a PhD in psychology. His team at Monell studies the science of smell in humans and how it affects our behavior.

As both a Swede and an expert in the science of olfaction, Johan Lundström told me he has never, ever, smelled anything as horrid, as putrid, as sickening as the Swedish canned fish delicacy (atrocity?), *surströmming*.

"Its odor is the most foul I've ever experienced. And I've been working in olfaction for ten years," Lundström says.

In ancient Sweden, salt was used to preserve everything, including fish, the mainstay of the Swedish diet. When the herring catch was very large and the supply of salt—expensive in ancient times—was very small, the Swedes had to get creative. Instead of curing the fish with salt, they resorted to another method to preserve an excellent harvest. They put the fish in cans along with water and just enough salt to control (but not completely halt) the growth of microorganisms, sealed the cans, and put them into storage. Because the amount of salt in the brine wasn't high enough to completely stop all microbial growth, what happened inside the can was a type of oxygen-free—anaerobic—fermentation. In food processing we go to great lengths to avoid anaerobic fermentation. Usually, this type of spoilage gives off signals so you know not to consume the food. Sometimes the gases that develop in cans of surströmming are so strong that the cans buckle and bulge: a clear warning signal if ever there was one. Yet in Sweden this only increases the value of the putrid fish inside.

Lundström correctly refers to the above process as *rotting*. This rotten herring product is still being made in modern Sweden, although you have to be pretty committed to eat it. Swedish law prevents apartment renters from opening the putrid cans inside their homes because the stench is almost impossible to remove from a building. It's also prohibited on some international flights. The preferred method for getting the rotted fish out of the can is to hold it under water while opening it, so the offensive aromas that escape are largely lost in the water, as opposed to being volatized into the air. This method also acts to control the stinky contents from spraying everyone in the near vicinity of the can, as a lot of pressure builds up inside it during fermentation. Years ago, Lundström took his Canadian girlfriend to Sweden, and in an effort to share his culture with her, he and a few friends threw a surströmming party. Outside, of course. Out came the pièce de résistance: the rotted, canned, fermented herring. When the uninitiated foreigner got a whiff of it, she vomited.

Ever the curious eater, I went online to order surströmming from Sweden, but I was unsuccessful every time I tried. The U.S. Customs and Border Protection doesn't look favorably upon bulging metal cans with foreign words on the label. So I asked Lundström to tell me what it tastes like.

"When you put it in your mouth, the odor is completely different," says Lundström, describing the taste when surströmming is served classically, as opposed to the nauseating smell.

It's a little bit sour. You have this very fresh taste of onion, the warm boiled potato, and the sour note of the herring on the crispy bread. That together is extremely nice. There's this complex difference between the retronasal odor and the orthonasal smell you detect through your nose before you eat it. And no one really knows what this is.

This conundrum also fascinates Linda Bartoshuk. At her Center for Smell and Taste, Bartoshuk has experienced the same distinction between aromas that are detected outside the mouth versus inside.

"You take the same molecule. You sniff it, you like it. You put it in your mouth and you don't like it," she says, explaining a phenomenon that happens with some food.

Bartoshuk told me about a patient who came into the Center for Smell and Taste after injuring her tongue in a cringe-inducing accident months earlier. The woman had opened a metal can, stuck her tongue into it and licked the sharp inside, slashing the nerves in her tongue in the process. (I told you it was cringe-worthy.) Bartoshuk expected her to complain of a loss of *taste*, since she'd injured her tongue, where we *taste* food. But she visited the clinic months after the accident because she was being tortured by the smell of her mother-in-law's lasagna.

Before the accident, the patient used to love the homemade version of this Italian specialty as prepared by her husband's mother. After the accident, she would still salivate at the heady aroma of Italian cheeses and tomato sauce bubbling in the oven until gooey-crisp. The problem was that when she finally sat down at the table to eat it, it tasted like cardboard.

Bartoshuk's first reaction was that this woman was lying in order to get some kind of insurance settlement. As a good practitioner of science, though, she conducted an experiment on herself to try to replicate the injury's effect. She ate a bite of a milk chocolate Hershey bar and noted the sensations. Delicious. Creamy. Chocolaty. A little bit sour. Roasty. Then she anesthetized her tongue—which she could do since the Center is affiliated with the school of dentistry—but didn't do anything to alter her sense of smell, because she was trying to duplicate the medical condition of the woman who had a slashed nerve in her tongue but an intact sense of smell. With her leaden tongue, Bartoshuk then ate a piece of that same chocolate bar. She was shocked:

> It wasn't chocolate anymore. The retronasal olfaction should have been just like normal. It should have gone right up, back, and into my nose, no problem. Here's what I think goes on. The brain looks for a cue, to tell itself whether that particular odor came in through the nostrils or from the mouth. The cue is: When you sniff, the brain knows the smell

came in through your nostrils. When you're chewing, swallowing, and getting taste and touch in your mouth, the brain knows it's coming from your mouth. It sends that olfactory information to different parts of the brain for processing, depending on that cue.

Now what happened to this woman with the cut nerves, she didn't get *either* cue. She wasn't sniffing it when she put the lasagna in her mouth. But the brain wasn't getting the clue from the mouth either, since the nerves had been cut, so it threw the information away. Somehow the brain needs the information from the taste system to process retronasal olfaction [mouth-smelling].

Normal people with functioning taste and smell senses wildly underappreciate mouth-smelling. Scientists do not. This is a hot area of research. You can experience mouth-smelling without tasting, in the exercise at the end of this chapter.

Breathing Scented Air

Molecular gastronomy, the high-concept trend in cooking that took off sometime after the turn of this century, uses food science to create mind-blowing food that is visually beautiful or arresting and presented in challenging new forms, tastes, and textures. It is often also delicious. Chefs describe their dishes as being freeze-dried, foamed, or sealed under vacuum and cooked in a controlled-temperature water bath (sous vide); these are some of the same techniques that we use at Mattson. Many of the ingredients that make molecular gastronomy possible, as well as the techniques, are common to our business. For example, at Mattson we've been using hydrocolloids (compounds used in food to manipulate texture and viscosity) as functional ingredients for more than three decades. In professional food development, we strive to balance culinary art with science, but unlike molecular gastronomy restaurants, we try to minimize the visibility of the science. Consumers generally feel better about eating food that's less rather than more processed, so we use the minimum amount of processing necessary. And we certainly don't market to customers how we do it. In fact, the less consumers read about the science that makes their food safe, extends its life on the shelf, or delivers certain nutritional benefits, the better. The fact that restaurants like Alinea in Chicago and wd~50 in New York put food-processing science in the forefront makes me just the teensiest bit uncomfortable, but restaurant diners have been eating it up.

What I really love about molecular gastronomy is the way it pushes boundaries to make you slow down and more carefully and thoughtfully consider the plate of food in front of you—and perhaps other meals you eat from that point forward.

Years ago, I enjoyed a twenty-three-course meal at Alinea restaurant with some girlfriends. At about course number seven, twelve, or maybe even seventeen, our servers brought three small pillows to our table, gingerly put one in front of each of us where our place mats should have been, and instructed us not to touch. Food runners from the kitchen arrived shortly thereafter with three piping-hot plates of lamb, which they set carefully atop our pillows. The waiter explained, "Your places are set with pillows filled with coffee-scented air. As you cut the lamb into bites, the pressure on the plate will force air out of the pillow, surrounding you in the aroma of coffee, which we've paired with the lamb for a totally new taste experience. Enjoy."

The three of us looked at each other and burst into giggles. When we recovered, we attempted to eat the course. I later heard Achatz interviewed about this technique, explaining that his goal was to add another sensory counterpoint to the table. I'm all in favor of this, but his technique gave an additional aroma to the environment, not the food. The difference is nose-smelling versus mouth-smelling.

When I tell Bartoshuk about this high-concept dish she cracks up, too. She says that we humans are extremely good at separating aroma in the environment from what we experience internally through mouth-smelling. As an example, she cites a big holiday meal at a table that someone has graced with a fragrant bouquet of flowers. The floral aroma that wafts from the flowers doesn't interfere with your experience of eating, she says. In other words, the flowers don't make the turkey taste like roses, or the mashed potatoes taste like calla lilies. That's because we only nose-smell the flowers. The turkey flavor we get from both nose- and mouth-smelling. We smell the flowers, but we savor the food. The coffee-scented air didn't change the flavor of the food, because we are too good at distinguishing aromas that stay outside our body from those that we experience inside the body. The fusion of taste and aroma is hard to separate, but artificially adding nose-smelling while eating food is not the same as mixing taste and aroma together in your mouth.

Still, the smell of calla lilies on a table can dredge up memories, emotions, and associations with the fragrant flower, as Achatz was trying to do with the coffee-scented air to evoke a Fragrant Flashback. He's also done this by burning oak leaves to evoke the crisp air of autumn that smells of fragrant, fallen-leaf bonfires.

Humans are spectacularly sensitive to smell. In some ways, it is our most finely tuned sense. We can detect some smells at the level of parts per billion—this is akin to a few drops of liquid in an Olympic-size swimming pool. If you set up an experiment where two pools were sealed off from each other, one containing three drops of the chemical ethyl mercaptan, the other without, most people would be able to tell which pool had the chemical in it solely by the skunky smell that the chemical gives off. Ethyl mercaptan is so strong that it is often used as a warning agent in gases such as butane and propane so that we are alerted to even tiny gas leaks. Obviously, not every aroma is this strong, but the example does illustrate the sensitivity of our sense of smell. In contrast, our sense of taste is only sensitive down to a few parts per hundred.

It would stand to reason, then, that we'd be really good at smelling something with our eyes closed and naming the smell. But we're not. Research has proved again and again that we bomb at this task. It's difficult to pull the names of aromas out of thin air, mostly because we just don't learn this skill. Food aromas come in through our mouth accompanied by taste, texture, temperature, a visual glance, and, most important, context. Separating these sensations and then putting a name to the smell piece of the puzzle isn't naturally in our repertoire of skills.

On the Bravo television show *Top Chef*, cheftestants are put to the quick-fire challenge "Name That Ingredient," which shows just how hard this is—even for highly experienced, classically trained chefs. In a "Master Chefs" episode, professional chefs Rick Moonen, Susur Lee, and Jonathan Waxman faced off over a Thai green curry sauce with twenty-nine ingredients. Thai green curry sauce is incredibly aromatic, full of exotic ingredients such as galangal, fermented fish sauce, kaffir lime leaves, and lemongrass. I would have bet that a simple sniff would have given them all the information they would need, yet they were allowed to taste the sauce as well, giving them more assistance from their sense of taste and retronasal olfaction. Then they had to name an ingredient, one at a time. The first chef to name something that was not in the sauce would be eliminated.

Jonathan Waxman knew right away what he was tasting. "I have a fantastic wine palate, so I think my sauce palate is pretty good. Obviously, it's a Thai curry sauce. That was kind of easy," he told the camera. Yet Waxman was eliminated from the challenge when he named butter as an ingredient, after only four other ingredients had been named (coconut milk, garlic, lemongrass, and kaffir lime leaf). Waxman blamed his slip-up on trying too hard to make a bold proclamation, but the fact that he missed on the fifth ingredient is more than just a bit interesting.

It just so happens that our ability to discriminate odors in a mixture (think: sauce, dressing, soup) is actually limited to four. This statistic is astounding when you consider how many ingredients go into most recipes. This task seems so simple on the surface; yet even master chefs fail regularly at it.

But all is not lost! Research has also proved that learning and practice can improve our ability to discriminate aromas. Wine tasters are better able to put a name to what they smell in a wine goblet than nonprofessionals are. Ordinary people who work in perfume stores are much better at correctly putting a name on a smell than people who don't work with perfume.

My own professional experience has also borne this out. If you want to improve your ability to "name that food" by smell alone, I suggest you do my Spice Rack Aroma Challenge exercise. Being able to name a spice by smell alone seems easy—especially if you cook with spices often. Yet this task frustrated me for months until I practiced over and again. Now I can tell clove from cardamom, thyme from marjoram. But if you mix together more than five or six of the spices I know by smell, you'll likely stump me after the first three or four. On the plus side, though, you'll have given me a warm-and-fuzzy Fragrant Flashback to my childhood in Baltimore.

If you're new to this, a good way to drive yourself completely crazy is to sniff a spice and—from that alone—try to identify what it is. Instead, try to identify the aroma category that it falls into. For example, is it a warm spice, such as the type of ingredient you use to make pumpkin or apple pie? If yes, then what spices fall into this category? A few that come to mind are cinnamon, nutmeg, clove, and allspice. Is the smell musky, sweaty, or animal-y? These are the aromas I get from cumin or white pepper. Is it grassy? Dried dill and parsley smell of lawn cuttings to me. Identifying the aroma category will allow you to narrow down the list of choices from the entire universe of herbs and spices to a subset of them. This way, you'll drive yourself only semicrazy.

Since identifying an aroma out of thin air is so difficult, how do you test patients who complain of smell loss? If you were to give them a sniff of something, how would you know whether they could smell it or not? Most people find this task difficult when their senses of smell and taste are working! Instead, doctors generally use some form of a multiple-choice test.

Others use an olfactometer, a special piece of equipment that puffs volatile aroma molecules into a tube or cylinder that you lean in to sniff. A computer screen or paper gives multiple choices for what you smell. By varying the concentration levels of the aromas, an olfactometer can measure not just whether you can smell or not, but how sensitive your sense of smell is.

An annual checkup at the doctor's office may include an eye and ear exam, but it rarely includes a test of your ability to smell. This is a terrible oversight, since loss of smell can put your health in danger. Loss of smell changes how palatable your food is, often affecting how much and what you eat. It also puts you at other risks, too, if you can't smell gas leaks, the smoke from a smoldering fire, or food that has spoiled.

Doctors don't even need an expensive piece of equipment for conducting smell tests. There's a standard paper test in the industry called the UPSIT: the University of Pennsylvania Smell Identification Test. It's a handy little booklet of scratch-and-sniff questions that your doctor can administer in minutes. You scratch a brown patch of microencapsulated aroma and fill in the circle next to your answer, as you did on the SATs. Your choices are compared against the answer key for scoring. There's even a way the test can determine if you are purposefully making the wrong choices. This is to identify people who visit their doctor hoping for an insurance settlement, falsely claiming they've lost their sense of smell.

The Brief Smell Identification Test and the University of Pennsylvania Smell Identification Tests rely on multiple-choice questions, because humans are so bad at naming an odor without a hint. A typical scratch-and-sniff question looks like this.

This odor smells most like:

a.   fruit

b.   cinnamon

c.   woody

d.   coconut

You can test yourself by ordering a kit from www.barbstuckey.com.

What Is Flavor?

Harry Klee thinks supermarket tomatoes suck. He is a professor at the University of Florida Horticultural Sciences Department, where his lab is working to understand the chemical and genetic makeup of flavor in fruits and vegetables. The research he is most excited about these days is called The Tomato Project. Klee is motivated by the frequency with which we consumers purchase beautiful, intensely colored tomatoes that leave us heartbroken upon the first bite.

Said Klee, "Since the 1980s, we've known that the tomatoes being sold in the store were crap. They suck."

Alice Waters, the doyenne of local and seasonal cuisine, might say that this is because the U.S. industrial food system is all wrong and people should eat only local, seasonal produce. Her organization, The Edible Schoolyard, puts gardening into the public school curriculum so that students get to grow their own tomatoes from heirloom seeds. This type of tomato ripens into a delicate, juice-packed monster that tastes better in part because it's grown from heirloom seeds, in part because it's eaten shortly after being picked. For farmers to sell heirloom produce, they have to carefully transport their flavorful, water balloon–like tomatoes only a few miles from the farm. They can sell the tomatoes in small though expensive quantities during only a few months of the year. For these few months, the consumer is happier and healthier, with a tomato that tastes like a tomato, and the farmer has a few more dollars in his pocket. Yet most of the produce in the United States is shipped from where it was grown to where it will be consumed by airline or truck, a process that would destroy heirloom tomatoes.

Klee believes that, by understanding what makes those heirloom tomatoes taste so good, we can reengineer them to survive being picked in Florida and shipped by truck to Minnesota in February without sacrificing flavor. Don't let this scare you. The technique Klee is using is not genetic engineering. It's conventional crop breeding, which farmers have been doing for hundreds of years. To breed the best tomato, though, he needs to know every single chemical molecule that makes up tomato flavor and then figure out which of the molecules are considered delicious. But what *is* tomato flavor?

"Tomato flavor basically is sugars, acids, and volatiles," says Klee. "You have to have sugar. You have to have acid. The balance of those two is very critical to good flavor. You have to have the foundation of sugars and acids, but we believe the volatiles are where the action is. You can't take sugars and acids and reconstitute tomato flavor."

Sensory Snack

Our perception of freshness is largely driven by acidity. Yet here, again, volatiles are where the action is! A long-simmered or jarred sauce may have plenty of brightening acidity, but if it's cooked so long that all the volatile top notes have evaporated off, the sauce will lose its freshness. This is also true for soups, salsas, fruit . . . everything. It's not just that eating fresh food is good for you, it actually tastes better because it contains more volatiles.

Capturing Aroma

To reconstitute tomato flavor, you could start with a piece of equipment called a gas chromatograph mass spectrometer (GC-MS), which is what many fragrance companies use to map aromas. This machine reads the volatile aromas of a food. Feed this mysterious black box a ground-up fresh tomato and it spits out the list of volatiles that are present in the sample, along with a spiky diagram of the strength of each, regardless of whether the volatile smells are at concentration levels humans can smell or not. Yet holding this list in your hand is like reading the ingredient statement on a can of Coca-Cola. Knowing that it contains carbonated water, sugar, caramel color, phosphoric acid, natural flavors, and caffeine isn't enough to let you whip up something that tastes like Coke. Nor does it tell you which ingredients are in the recipe for what reasons. A reading from a GC-MS gives you this kind of information, but it's never the whole story. The results of the black-box GC-MS test are just as useless for knowing which volatile components of the tomato make us like it. This is what Klee is trying to figure out.

While there are only five Basic Tastes, the number of aromas that exist is far larger. There are about 400 volatile aromas in a tomato, many of them at levels too low for humans to detect. There are only about fifteen volatile chemicals that give a tomato its characteristic flavor.

"The interesting thing to me about tomatoes is that not one of those chemicals would really jump out at you as tomato flavor if you smelled it," says Klee. This is not true of every food. For example, cinnamon has one dominant, signature volatile compound (cinnaminic aldehyde); the same is true of cloves (eugenol) and butter (diacetyl).

"Tomatoes don't have that," says Klee. "It's the sum of the parts that gives you tomato aroma." The list of the volatile compound "parts" of a tomato that humans can smell is shown below. For example, there are green, grassy notes that smell a lot like the cut grass that lies atop your yard after you mow. And they're very important to tomato flavor. In fact, says Klee, those tomatoes you buy on the stem are taking advantage of this fact. "It's an incredible gimmick. People buy the cluster tomatoes. In the store they pick up the cluster and they smell it and they say, 'Oooh, that smells good.' They're not smelling the tomato. They're actually smelling the vine."

The Naturally Occurring Volatile Compounds
That Make a Tomato Savor Like a Tomato

Aroma	Volatile Compound	Concentration
Tomato/green	*cis*-3-hexenal	12,000
Green/grassy	hexenal	3,100
Nutty/fruity	2-phenylethanol	1,900
Fruity floral/green	1-penten-3-one	520
Earthy/musty	3-methylbutanol	380
Green	trans-2-hexenal	270
Green	*cis*-3-hexenol	150
Fruity/floral	6-methyl-5-hepten-2-one	130
Green	*trans*-2-heptenal	60
Wintergreen	methyl salicylate	48
Tomato vine	2-isobutylthiazole	36
Musty	2+3-methylbutanal	27
Musty, earthy	1-nitro-2-phenylethane	17
Floral/alcohol	phenylacetaldehyde	15
Fruity/floral	β-ionone	4
Fruity	β-damascenone	1

From The Tomato Project website: hos.ufl.edu/kleeweb/flavorresearch.html

My fiancé, Roger, hates fresh tomatoes; I can't imagine life without them. It's really difficult for him to articulate what he doesn't like about the tomato. His answer, "There's a certain flavor . . . ," is accompanied by a scrunched-up face. Yet he adores foods made from cooked tomatoes: pasta sauce, tomato soup, ketchup. The difference between the two is that the cooking (or processing) of tomatoes flashes off the volatile aroma compounds in the tomatoes that Roger finds objectionable. Cooked tomatoes don't have those green, grassy, earthy top notes, defined as compounds with low molecular weight that are really sensitive to heat. Top notes are the aromas you detect most strongly in a food. Because they are light, they evaporate quickly. This phenomenon explains why freshly squeezed orange juice savors better than pasteurized. The low molecular weight volatiles in the juice are lost from the heat of pasteurization. They are the compounds that give it that fresh flavor.

The background notes in tomatoes withstand cooking because they're heavier. The weight of these molecules makes them less sensitive to heat. I don't know if it's the trans-2-hexanal, the β-ionone, or the 3-methylbutanol that offends Roger. All I know is that four to five minutes in a sauté pan are enough to get rid of it.

Now, let me recommend a real-world experiment that my father—not a scientist—taught me at age five. In the middle of August, go to the best grocery store or farmers' market in your neighborhood. Find an heirloom tomato that's heavy for its size, plump, full of juice, but not mushy. When you get home, put a paper towel in the neck of your shirt as a bib. Lean over the sink, wash the tomato. Sprinkle it with finely ground sea salt. Then, holding it like an apple, sink your teeth in. See if you can detect the phenylacetaldehyde. No? Good. Now forget about volatiles and enjoy one of Mother Nature's most perfect foods.

Taste: How Heating Affects Volatiles

YOU WILL NEED
>   Masking tape and marker
>   2 glass measuring cups
>   2 whole lemons for each person tasting
>   Juicer
>   A medium bowl
>   Plastic wrap
>   A small nonreactive saucepan
>   Saltine crackers for cleansing your palate

DIRECTIONS
>   1. Using masking tape, label the glass measuring cups.
>   Measuring cup 1: Fresh Juice. Measuring cup 2: Boiled Juice.
>   2. Juice all the lemons into the bowl.
>   3. Pour the juice into the measuring cups until you have
>   separated it into two equal parts.

4. Cover the measuring cup labeled "Fresh Juice" with plastic wrap and put it in the refrigerator.
5. Pour the remaining juice into the saucepan. Heat on high until the juice starts to boil. Boil on medium-high heat for 3 minutes.
6. Remove the juice from the heat and pour into glass measuring cup labeled "Boiled Juice."
7. Chill the boiled juice to the same temperature as the fresh cup, about 2 hours. Make sure the two cups are the same temperature.

TASTE

8. Taste the fresh juice first.
9. Eat a cracker and drink some water to cleanse your palate.
10. Taste the boiled juice second.

OBSERVE

1. When you taste the fresh juice, you will experience two— perhaps three—Basic Tastes. You'll definitely taste sweet and sour (mostly sour) with perhaps a tiny bit of bitter, depending on the lemon and how much of the pith and rind you squeezed.
2. After you've noticed the tastes, focus on the aromas. Freshly squeezed lemon juice has some brilliant, crisp, sharp aroma top notes that give it its signature flavor.
3. When you taste the boiled juice, you'll notice the same basic tastes. But when you focus on the smell in the boiled juice, you'll notice that something's missing. It's not as aromatic as the fresh juice. This is because boiling the juice liberated the lighter compounds and they evaporated off into the air. The resulting juice thus contains fewer aroma compounds.
4. The process of boiling juice is similar to what happens when juice (or any liquid) is pasteurized, a process that uses heat to kill bacteria. It allows juice to stay fresher longer. The heat

kills dangerous microbes as well as lovely, subtle aromatic top notes. Next time you taste a glass of orange juice at a restaurant, see if you can detect those subtle top notes, which might indicate whether the restaurant is squeezing its own juice. Or not!

Taste: Nose-Smelling Versus Mouth-Smelling

YOU WILL NEED

    1 bottle of nectar-like juice such as pear, peach, or mango, at
        room temperature
    3 glasses or plastic cups with matching lids
    A roll of plastic wrap (if you are using glasses)
    1 stick butter, brought to room temperature naturally
        (do not melt using heat!)
    1 jar peanut butter, at room temperature
    Sampling spoons and cups
    3 bendable straws for each person tasting
    Paper and pens or pencils

DIRECTIONS

PREPARE THE STIMULI

1. Pour ½ cup juice into one of the glasses. Cover the top tightly with plastic wrap or a snap-on lid.
2. Put 2 tablespoons butter into one of the remaining glasses. Cover the top tightly with plastic wrap or a snap-on lid.
3. Put 2 tablespoons peanut butter into the last glass. Cover the top tightly with plastic wrap or a snap-on lid.
4. Set the glasses aside for about an hour so that the volatile aromas of the juice, butter, and peanut butter fill the empty space in the glass. This empty space is called the headspace.

The short end of the straw goes in the cup but shouldn't touch the food.

Put your lips around the straw and breathe in and out to experience mouth-smelling without nose-smelling

Set the remaining juice, butter, and peanut butter out so that they are easily accessible for sampling but out of smelling range of the covered glasses.

SMELL AND DISCUSS
- Using the short end of a straw, the first person should puncture the plastic wrap or lid with it, but do not let the straw touch the aroma stimulus. Make sure it stays in the headspace. You want to sample the air, not the food or drink.
- Close the lips around the straw and breathe in and out. This is the retronasal smell, or mouth-smell, of the stimulus. Allow each taster to mouth-smell the samples (using clean straws, if they want).
- Now, smell the stimuli as you normally would, using your nostrils.
- Finally, taste the stimuli.
- Discuss your experience.

Taste: Spice Rack Aroma Challenge

This is an ongoing training exercise that will take you a few weeks, maybe even months, to master.

YOU WILL NEED

10 to 15 spices and herbs that you use on a regular basis
Masking tape
Marker
Paper

DIRECTIONS

1. Using masking tape, put the names of the ingredients on the bottoms of the containers.
2. Cover the container labels with paper so that you can't tell what's what.
3. Quiz yourself by randomly picking up a jar, opening it, and smelling. Try to identify the contents without looking at the name. *It's best to do this with a partner, as the temptation to cheat will be overwhelming! You will not learn if you look at the name of the spice first and sniff second. You must sniff first and guess second.*
4. At first, identifying spices by smell will be extraordinarily hard. You'd think you'd know the smell of ingredients you use regularly! The problem is that you're never forced to identify them without other contextual clues.
5. If you find this difficult, ask yourself this series of questions:

   a. Is this an ingredient you use regularly? If no, you're in trouble! If yes, proceed.
   b. Is this a familiar smell? Yes or no?
   c. If it's familiar, try to associate the smell with some emotion it provokes. For example, is it something you cook with often or not? If yes, that helps. If no, that gives you a clue as well. Is this something you've smelled at your mother's house? Friends' houses? That Indian restaurant down the street?
   d. Try to identify what category this ingredient belongs to. Does it smell citrus-y, musky, grassy, or warm? If you can

identify the category of aroma, that will help narrow it down.

e. What do you think it is? Take a guess before you look at the name. This is the only way you'll learn!

Grassy	Pungent	Warm	Licorice-y	Citrus-y	Musky, Dirty	Smoky	Floral
Basil (dried)	Pepper	Cinnamon	Basil	Ground coriander	Cumin	Red pepper flakes	Lavender
Oregano (dried)	Mustard	Clove	Fennel	Bay leaf	White pepper	Chipotle chile powder	Herbes de Provence
Dill weed (dried)	White pepper	Allspice	Caraway seeds			Smoked paprika	
Sage (dried)	Ginger	Cardamom	Anise seed			Cayenne pepper	
Parsley (dried)		Nutmeg					

Source: L. Barthomeuf, S. Rosset, and S. Droit-Volet, *Emotion and Food*.

Taste: Appreciating Mouth-Smelling

This exercise will illustrate how much longer flavors last when you keep them in your mouth and employ mouth-smelling to extend the flavor experience.

YOU WILL NEED:
  Prepared gelatin dessert (such as Jell-O) for everyone
    tasting
  1 spoon for each taster
  Saltine crackers and water for cleansing your palate
  Pens and paper for writing down scores

INSTRUCTIONS:

1. For each sample, the goal is to count how long the flavor of the gelatin lasts in your mouth. Start counting from the time the gelatin enters your mouth to the time the flavor is completely extinguished.
2. Toss a spoonful of gelatin to the back of the mouth and swallow it as fast as possible. The goal is to get it down quickly, without much chewing.
3. Count how long the flavor of the gelatin lasts.
4. Eat a saltine and drink some water to cleanse your palate.
5. Place a spoonful of the gelatin on the center of the tongue, close the mouth, and keep it there as long as possible, being sure to breathe in and out as long as it's in the mouth. Move it around while you keep breathing. The goal is to hold it in your mouth as long as possible before swallowing.
6. Count how long the flavor of the gelatin lasts.

DISCUSS:

- How much longer did the flavor of the gelatin you held in your mouth last?
- Imagine if you could get more flavor out of everything you eat this simply!
- All it takes is being more careful to hold food in your mouth, move it around to touch all the taste buds, and breathe continuously so that you can experience the volatile aromas.

3

Touch

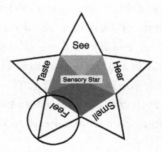

Most people wildly underappreciate how much their sense of touch influences what they eat. Cameron Fredman, a litigator born and bred in Los Angeles, considers himself the unofficial spokesman for the appreciation of food texture.

Fredman derives great pleasure from food even though he's never in his thirty-some years smelled or tasted it. He enjoys cooking and eats out often. Yet he is a connoisseur of food texture by default. That's because Fredman was born with ageusia and anosmia: total absence of the senses of taste and smell. He will never truly know what food flavor is: the combination of the taste, smell, and texture. Fredman's entire experience of flavor is solely texture.

As persuasively as lawyers defend their clients in court, Fredman finds himself defending the importance of texture in food. He talks about food texture the way chefs talk about flavor. "When discussing taste with people, I'm often pitching texture. I'm arguing that a lot of what they like about food is *texture*, which gets credited to the taste of the food. Good food has good texture. Poorly prepared food has poor texture."

Fredman waxes poetic about the layers and complexity of food texture. In describing one of his favorite foods, he says, "With sushi you have many bite-size, uniquely prepared varieties of texture." He describes good sushi as being made with properly cooked, firm rice placed next to taut-textured, meaty fish. In contrast, poorly made sushi is a much more homogeneous textural experience.

Most people think rice cakes savor like Styrofoam (almost literally) because they lack a strong signature taste or aroma. But to Fredman, who relies solely on texture, there's nothing missing from the snack disks that many people consider punitive diet food. "Rice cakes are just sort of interesting, texturally, if you stop and think about the way that the rice breaks off from the rest of it," he says. "I think it's more interesting than other kinds of crackers."

One problem with Fredman's reliance on food texture is that he is quick to lose his appetite when the texture of a food doesn't match his expectations. A walnut hidden in a scoop of fresh yogurt might trigger a gag reflex. Most people would taste the bitterness of the walnut skin, then experience the mouth-smelling that would signal *walnut*. Fredman doesn't get those taste and smell cues. To him, a walnut in his yogurt could just as easily be a severed body part or a piece of plastic. He usually gags out foreign textures, which is his own way of dealing with potentially dangerous food. He also adheres strictly to the expiration dates on packaged foods because he's unable to stick his nose into a carton of milk to smell if it's fresh. He relies heavily on visual spoilage cues, such as the mold that develops on cheese or the slime that coats a rotten vegetable. He's not always successful at avoiding harm, though. When the texture of a food is correct, but the taste or smell is not, there's a chance that dangerous things can get past his gullet. As on the sweltering night he spent in Venice Beach, California, without air conditioning.

All Liquid "Tastes" Like Water

Fredman was sweaty, tired, dehydrated, and uncomfortable in the dark. His girlfriend always had a bottle of water handy, so he asked her where it was. While they were lying in bed, she told him her bottle of water was on the table. He sat up, reached around until he felt it, then grabbed the bottle and chugged. After a few swallows, Fredman realized the resulting dizziness and nausea he was experiencing wasn't from sitting upright quickly in the heat; they were from whatever the hell it was he was drinking. When his girlfriend turned on the light, he saw immediately from the horrified look on her face that something was terribly wrong. Then he looked at the bottle he was holding. It read: PINE-SOL.

Fredman lacks the two sensory systems that would alert most people to the danger of drinking Pine-Sol: smell and taste. Without another sensory element to differentiate it from water, Fredman was drinking blind. He sensed no textural difference between water and Pine-Sol. There was no auditory difference. And he couldn't see the color or label because it was night and the lights were out.

Yet that didn't mean his body accepted the liquid as water. He vomited for hours, but the dispatcher at the end of the 911 line told them this was a good sign. Once it was clear that Fredman would live through this frightening experience, his girlfriend couldn't help but find the humor in it. "You should vomit on the bathroom floor," she suggested of his sudsy discharge, "and wipe it up between bursts." He wasn't just a boyfriend; he was a floor wax, too.

Cameron Fredman's pleasure in eating, in which he uses only his senses of sight, sound, and touch, shows us that we can—and do—derive great pleasure from food texture. Yet when we remember food experiences, it's rare that texture is at the top of our minds, except perhaps when something is extraordinarily high in fat. My mother goes giddy over the silken texture of the flan at her favorite Mexican restaurant. I've listened to Roger talk reverently about the texture of Humphry Slocombe Secret Breakfast ice cream, which he loves because it stays soft due to its high alcohol content, even when it's been stored at zero degrees. The bourbon paired with cornflakes in vanilla ice cream simulates a bracing, hair-of-the-dog breakfast. Not that I would know about that.

Sensitive Information

Imagine yourself feeling a fresh tomato. It's likely that you are envisioning yourself touching it with your hand. This makes sense, given that you have to use your hand to put a tomato into your mouth. It also makes sense because your hand is your most sensitive body part. The next most sensitive body parts are your lips and tongue. After touching the tomato with your hand, if you want to get the most feel from it, it would make the most sense to stick it in your mouth. Your lips and tongue have almost as many nerve endings as your fingertips.

The image below is a graphic representation of the sensitivity of touch, called the Sensory Homunculus. The size of each body part has been increased or decreased in accordance with how discriminating it is. Notice how big the hands, lips, and tongue are.

The Sensory Homunculus

Now imagine that a friend has bought a new cashmere sweater that she really loves. "It's so soft," she might say to you. "Go ahead, feel it." It's unlikely that you will lean in close, gather up a handful of her sweater, and stick it in your mouth. Yet this is exactly what babies do! At some point in your maturation, you learn that tonguing your friends' sweaters is socially unacceptable behavior, even though this action would help you get the most feel out of the sweater. Your incredibly powerful sense of touch is inextricably linked with what you savor.

Why the Monkey Is Chunky

One of the great American textural food experiences is the famously chunky superpremium ice creams available in every grocery store these days. But this was not always the case. In 1978, two men set out to make great ice cream with superhigh butterfat content. The more butterfat the ice cream has, the creamier, richer, and more luxurious its mouthfeel—or how it feels in your mouth.

It so happened that one of the founders of the dairy company suffered from one of the same conditions as Cameron Fredman, *anosmia*, the lack of the ability to smell. During the early days when the founders were developing the recipes themselves, the anosmic one used to push for more and more stuff in the ice cream: more chunks of chocolate, more walnuts, more brownie pieces, more cherries. He couldn't smell the fruity-almondy aroma of a cherry; the roasty, nutty aroma of chocolate; or the burnt-sugar flavor of the caramel. He wanted texture to replace that missing olfactory input. Loading up their ice cream with swirls, chunks, fruit, and other inclusions gave the founder a little compensation for what he was missing.

This combination of high-quality ice cream and a-chunk-in-every-bite texture would become Ben & Jerry's signature. Ben Cohen's anosmia was partly responsible for the iconic textural indulgence that even today marks his and his partner Jerry's ice creams. Many competitors have mimicked their style, but Ben & Jerry's is still the texture-lovers' ice cream choice, with varieties like Chubby Hubby: fudge-covered peanut butter-filled pretzels in vanilla malt ice cream with swirls of fudge and peanut butter.

Texture Is Touch

Texture is the part of flavor we experience through our sense of touch. A food experience becomes complete when you layer on the other senses of sight, hearing, taste, and smell. Good restaurateurs know how to stimulate all five.

Joshua Skenes, the flaky but fantastically brilliant chef who owns the lovely restaurant Saison in San Francisco, serves a multicourse tasting menu that challenges the diner to think differently about food pairings and flavors. When Saison opened, it was a shock to the senses and to conventional expectations. It was in a converted catering kitchen space, on a back alley in a neighborhood not known for fine dining, and you had to walk through the dishwashing station to get to the dozen or so wooden tables in the afterthought of a dining room. Eventually, Skenes's talent broke through and his customers wanted a setting that matched the hefty price tag he wasn't shy about charging and they weren't regretful about paying. When he decided to remodel his restaurant, he considered the sense of touch in a unique way: "You need great food, great service, great wine, great comfort. And comfort means everything. It means the materials you touch, the plates, the whole idea that the silverware was the right weight. We put throws on the back of the chairs," said Skenes.

This was not what I expected to hear when talking to a chef about texture. I went into the discussion with Skenes thinking that a chef's attention to the tactile sense is focused primarily on *food* texture. But of course it makes sense that physical comfort is part of the pampering experience of fine dining. Now that I've experienced it at Saison, I think that providing a blanket on the back of a restaurant chair should be standard restaurant procedure. Swaddling yourself in microfiber takes the idea of luxury dining to the next level, especially in San Francisco when a 40° summer night can chill you to the bones.

Tomato Talk

A Lexicon for the Textural Attributes of Fresh and Processed Tomatoes

Attribute	Definition
Fibrousness	How much stringiness do you detect?
Juiciness	How much liquid comes out when you chew?
Mealiness	Mealiness is like the Supreme Court's definition of pornography: you'll know it when you feel it. It's characterized by fine, soft particles that are not pleasant and are not supposed to be in a tomato.
Pulp amount	Pulp relates to solids: how much tomato "flesh" versus skin and liquid.
Skin awareness	Skin can be good or bad.
Seed awareness	Are seeds present?
Thickness	How thick is the skin? How thick is the juice? How thick are the fleshy walls of the tomato?
Viscosity	How quickly does the juice pour? How liquid is it?
Astringent	Is there a dry, puckering mouthfeel?
Metallic	Does it taste like tin cans or aluminum foil?

One word we use for texture in common English is *consistency*, as in "How is the consistency of that cream of tomato soup?" In this case, the word *consistency* means texture. But the word has another meaning, too. Consistency can mean the degree of conformity, such as "Roger's levelheaded behavior lends a steady consistency to our relationship." In this case, it means "a harmonious uniformity." You can see how this word could cause problems when it is used to evaluate food. Consider the sentence, "This batch of tomato soup is not as good as the last. I'm having trouble with the consistency." Does this mean that the soup maker is having trouble with the thickness or having trouble duplicating the recipe from batch to batch? It's unclear. To avoid this confusion, I prefer the term *viscosity* instead to describe texture.

Are You Feeling Bitter?

Most people think of astringency as a type of bitterness. This is not really the case. Many bitter or sour foods are also astringent, which is probably where the confusion comes from. But astringency is detected with the trigeminal or (touch) nerve. It is a texture. Some of the most common, well-known astringent compounds are tannins, which give red wine, coffee, and tea their tannic, drying mouthfeel. Another favorite astringent is alcohol, which is dehydrating. When alcohol dehydrates it, the tongue feels like it's been dried out. That drying feeling is astringency.

Tannins act on your tongue by making certain mucoproteins in your saliva poke out and puff up their chests. These proteins don't play well with other proteins, which results in a dustup in the playground of your mouth. This clash causes friction, which means that your saliva is less lubricating. The result is that your mouth feels dry. This is astringency from tannins.

In order to feel the astringency from tannins, you need only a red grape, the redder or darker the better. Be sure to choose a juicy one, rather than a firm one. Put the grape in your mouth and, without chewing too strenuously, mash the grape up against the roof of your mouth with your tongue and work it until you've entirely separated the skin from the juicy insides. Then swallow the sweet and sour innards of the grape, but keep the grape skin in your mouth. Now you're ready to experience the sensation of astringency. Put the grape skin on your back molars and start chewing. Keep chewing until you start to feel something that's drying or perhaps puckery. That is astringency. You can now taste the distinctiveness of astringency and understand why it is commonly confused with the Basic Taste bitter. Astringency is the tactile sensation you're feeling on your tongue.

Tannins are also known as *polyphenols*, which are healthful antioxidant compounds found in plants. Some tannins occur in red wine and come from the crushed skins of the grapes. They're also present in tea, coffee, pomegranates, and nuts. These foods are usually balanced with a counterpoint Basic Taste or two. The astringency in tea, for example, is often balanced with a sour squirt of lemon and/or a sweet spoonful of sugar. Red wine makers strive for the perfect balance of tannin, acidity, and sweetness. Since astringency results from a decrease in lubrication, it would make sense that a good way to balance tannic foods is with something slick, like fat. That's exactly what adding cream to coffee accomplishes. The classic pairing of red wine with steak taps into this as well. The fat in a piece of red meat is the perfect way to lubricate the tongue between sips of astringent, drying Sangiovese or cabernet sauvignon.

A Feel for Anatomy

You use your teeth to feel texture in several ways: to grip food, bite off pieces, and chew to reduce the size of the food until it is small enough to be swallowed. Another way your teeth help you detect food texture is through their nerve fibers. When you use your teeth to apply pressure to food, that pressure jiggles the teeth in their sockets ever so slightly, sending information along the nerve fibers to the central nervous system to identify food texture, often unconsciously.

If I say the word *muscle*, the first image that pops into your head is likely one of a curled bicep, bulging from the weight of a barbell, or some other similar image of an arm, leg, or torso. We almost never think about the tongue as a muscle, yet that's exactly what it is. You use the power and finesse of the muscles in your tongue to move food in the mouth just as you use the muscles in your arms to move bags, babies, books, and barbells.

Saliva greases the skids in identifying and appreciating food texture. Its lubrication allows the tongue to move the food around easily. Without saliva, we wouldn't be able to moisten food enough to form it into a ball (the yummy-sounding wad or bolus) to be swallowed. The saliva glands act as the irrigation system of the mouth, releasing moisture when it's needed and cutting it back when it's not. Salivary glands shut down for the night when we go to sleep, which is why we wake up with a dry mouth and bad breath. Because saliva isn't needed for irrigation during sleep, it doesn't refresh the mouth then as it routinely does when we're awake.

Swallowing is the final phase of food texture detection. The active part of swallowing happens when the tongue forms food into a bolus and passes it to the back of the mouth. Once it's there, throat muscles contract to aid in swallowing farther down your gullet. These automatic reflexes, surprisingly, don't rely on gravity. The reflexes that move food down the throat will work even if the muscles are anesthetized or if you are hanging upside down. This is why astronauts can successfully eat and drink in zero gravity.

The Sound of Texture

Our perception of food texture is influenced by the sound of the food we eat. If you put a potato chip into your mouth and don't hear a crunching sound, you won't need much more information to know that the chip is stale, wet, or undercooked. In one study, researchers wanted to see if they could influence consumers' perception of texture by manipulating the sounds they heard while eating. They chose to test potato chips and in order to measure the sound, they needed to ensure that every chip tested was exactly the same, an almost impossible task with regular potato chips. Their solution was a stroke of genius: they used Pringles, surely the most uniform potato chip on earth.

Once again, the subjects in the study were outfitted with headphones, as in the cross-modality experiment Linda Bartoshuk did when she tried to match the sweetness of a Coke to sound. Whereas Bartoshuk asked her participants to adjust the sound, here the researchers varied the volume for the participants. As the participants bit into Pringles—all perfectly identical in size and shape—the researchers changed the decibel level of the sounds they played. When the overall crunching sound of the chip was increased, the chips were rated significantly crisper and fresher. When it was lowered, the chips were rated more stale and soft.

Food Without Feeling

What would happen if you could keep all of your other senses working but you were able to remove texture from the equation? Could you even tell what you were eating? You remove the texture from an orange when you juice it, and you can certainly tell how orange juice savors, right?

Pureed Food	% of People Who Correctly Identify the Food		
	Normal-Weight College Students	Obese Subjects	Elderly Subjects
Salt (taste standard for salty)	89	94	89
Coffee (taste standard for bitter)	89	87	70
Apple	81	88	55
Fish	78	81	59
Strawberry	78	81	33
Pineapple	70	75	37
Corn	67	69	38
Sugar (taste standard for sweet)	63	88	57
Carrot	63	44	7
Celery	59	63	24
Tomato	52	69	69
Lemon (taste standard for sour)	52	25	24
Banana	41	69	24
Beef	41	50	28
Pear	41	44	33
Walnut	33	50	21
Broccoli	30	50	0
Rice	22	12	15
Potato	19	69	38
Green pepper	19	25	11
Pork	15	6	7
Cucumber	7	0	0
Cabbage	4	0	7

Less than half of the subjects were able to tell what many of the foods were without textural clues.

The Basic Taste standards were pureed with cornstarch and water until they were the same viscosity as the other foods.

Source: Susan S. Schiffman et al., "Application of Multidimensional Scaling to Ratings of Foods for Obese and Normal Weight Individuals," *Physiology & Behavior* 21: 417–22; and Susan Schiffman, "Food Recognition by the Elderly," *Journal of Gerontology* 32, no. 5 (1977): 586–92.

Researchers tested this theory by having three groups of test subjects (healthy college students, obese, and elderly) taste different foods without the normal textural cues. They gave the subjects samples of various foods, cooked until soft and pureed until they all had approximately the same mouthfeel (at least as close as the researchers could get; they had to adjust some of the foods with water). They also included four of the five Basic Tastes in their test: salt, coffee (for bitter), sugar (for sweet), and lemon (for sour), which they blended with cornstarch to match the mouthfeel of the pureed solid foods. The subjects were not allowed to see the foods. But they were allowed to nose-smell them before eating, and after they put the food in their mouths, they were allowed to swish it around to experience the taste and mouth-smell of whatever it was.

Without the textural clues that are normally present, more than half of the participants had trouble identifying familiar, everyday foods such as beef, pear, and broccoli. Some foods with flavors that are unmistakably strong and characteristic, such as green pepper (vegetal, green), were much more difficult to identify. Only 19 percent of healthy students could identify pureed green peppers. Thus it seems that for many foods, we need our sense of touch to know even the most basic information—namely, what it is we've just put in our mouth.

Texture Contrast

Pioneering food texture researcher Alina S. Szczesniak explained that our love of contrasting textures has a biological basis:

> All people begin by consuming liquids and (essentially) all people advance to biting and chewing of solids. The sequence, liquid before solid, and the contrast between the passive sensation of liquids . . . and the active ones of biting and chewing is fundamental and creates boundaries to texture experiences and combinations.

Good chefs go to great lengths to add texture contrast to their plates, utilizing four different approaches: within a meal, on the plate, within a complex food, and within a simple food.

Contrast *within a meal* refers to the varying textures that occur as a meal progresses. For example, a meal might start with the smooth, easy, comforting texture of a pureed cream of mushroom soup. This eases you into the meal with a food that requires very little energy to enjoy. You simply spoon it in, swish it around (if you want to get the most flavor out of it), and swallow. From there the meal might progress to a crisp, fresh romaine salad, tossed with a creamy dressing and topped with crunchy toasted nuts. By the time the main course arrives, you're warmed up enough, texture-wise, to take on the firm chew of a New York strip steak or toothsome lobster. When the savory courses are over, dessert is usually a fairly soft affair, again, which closes the parentheses of the meal as effortlessly as it began. Hard textures in a dessert course, such as the crisp flamed top of a crème brûlée, are usually reserved for garnishes, when it comes to dessert, lest the meal close with too much violence. Most diners prefer a soft landing.

Within the meal, there's also texture contrast *on the plate*. Let's take that New York strip steak, for example. The classic pairing of steak and mashed potatoes has (in part) to do with the fact that the soft, yielding potatoes—whipped with butter, cream, and, if you're exceedingly fortunate, crème fraîche—provide a foil for the tougher steak. Your jaw uses energy to chew the hunk of steak, and is subsequently rewarded with a forkful of soft potato that requires little more effort than pressing it against the roof of your mouth and swallowing. Effort, reward; effort, reward. A side of onion rings might gild the textural lily with firm yet crunchy onions inside a coat of crispy fried buttermilk batter.

Craig Stoll, chef-owner of the famed Delfina restaurant in San Francisco, layers texture into a complex dish by adding the same ingredient at different times as it's cooking. Stoll says:

> We do this one risotto with crab and radicchio. We add the radicchio at two stages. First, we throw it into the pot at the beginning with oil and garlic—a whole clove that we pull out. We'll sauté it and caramelize the radicchio a little bit to develop some sugars. And then we'll build the risotto into that. We'll take the rice and add it and then we'll cook it and work it and add the crab. Then we'll add more radicchio two-thirds or three-quarters of the way through the cooking so that you'll have more texture. You're layering in texture at a couple of different stages in the dish.

Sometimes getting a texture you aren't expecting is frightening, as it can be for Cameron Fredman, the attorney who cannot taste or smell, but sometimes it's a delight. Creative chefs such as Heston Blumenthal of The Fat Duck in England, Wylie Dufresne of wd~50 in New York City, and Grant Achatz of Alinea in Chicago like to rethink and reengineer familiar foods to surprise and delight diners. By aerating a food, they can change the texture of something that's usually hard, crunchy, or crispy into foam or "air" that melts quickly on the tongue. Aerating rich foods such as foie gras delivers the same intense flavor without the fill factor.

In Healdsburg, California, Cyrus Restaurant's chef-owner Douglas Keane serves a beer bubble that he makes with Racer 5 India Pale Ale, to which he added lemon juice, honey, salt, and sodium alginate. Just before serving it, he dropped it by the tablespoonful into a bath of gellan gum, water, and sodium hexametaphosphate, which formed a semisolid shell around the drop of beer. When I put it in my mouth, the bubble transformed instantly to liquid. Without Keane's manipulation and use of high-tech ingredients, the beer would have been a singular, familiar homogeneous liquid texture. With his artistry, the texture became something altogether different. As Szczesniak says, "The abrupt change from a solid to a liquid state is very pleasurable."

But don't think that this type of culinary skill (or pantry of six-syllable ingredients) is required to demonstrate texture contrast *within a complex food*. Simply reach for the nearest Oreo cookie. There's perhaps no better example of texture contrast within a food than the mouth-coating sweet and fatty creme filling tucked between two crunchy cocoa cookies.

Texture contrast *within a simple food* is the type of contrast that happens over time as the heat from your mouth changes a food's texture, without your needing to apply force or energy. As all chocoholics know, you can simply place a piece of good-quality chocolate on your tongue, close your lips, and revel in the blissful change of texture from solid to liquid: jaw, tongue, and teeth are not required. Sucking on a piece of hard candy or an ice cube results in the same type of temporal (time-related) texture contrast, from solid to liquid, yet these examples are obviously inferior to chocolate if you want to demonstrate the effect quickly (and indulgently).

Rutgers's Paul Breslin believes that we innately enjoy this type of solid-to-liquid texture change because it usually indicates the presence of fat, which is an incredibly efficient source of calories—exactly what you need when you're hunting and gathering. In fact, fat is more than twice as calorically dense as carbohydrates and proteins. Of course, we don't need those extra calories when we're hunting for a parking spot closer to the mall entrance or gathering dust on our couches each night. Yet this evolutionary reaction to fat is still with us. If you concentrate on how fatty foods like butter, cheese, and ice cream behave in your mouth, you can savor how pleasurable that shift—from hard or frozen to soft and creamy—feels. There's absolutely nothing like it.

Spit Genes

American-style chocolate pudding—the kind of dessert that Jell-O brand makes—is made from milk cooked together with cocoa, sugar, and starch until it thickens up. The key thing about this type of pudding is that it's the starch that gives it its texture.

In her lab at the Monell Chemical Senses Center, Abigail Mandel, a postdoctoral fellow from 2009–2011, put a dab of something magical into a texture-reading machine along with the pudding. When this amazing substance came into contact with the starch in the pudding, it "pulverized" it, said colleague Paul Breslin, meaning that it almost instantly changed the pudding from a thick, viscous texture into something thin and almost watery. The something they added to the pudding was spit. That's right: human saliva.

Before we go any further, I should clarify how they harvested the saliva for use in their experiments—because that was the first thing I wanted to know.

"We just have people drool into a cup," said Breslin, without a shred of humor in his voice. He went on to explain that this finding gives us enormous insight into why we like the foods we like. As with so many things, it goes back to genetics.

There's a gene that determines how much of a substance called *amylase* your saliva contains. The more copies of this gene you have, the more pulverizing amylase you have in your saliva. If you have lots of copies of the gene, and hence lots of amylase, you are going to notice a texture change in pudding much more quickly, and perhaps intensely, than someone who doesn't. Because texture contrast within a food is desirable, you may get more pleasure out of thick, starchy foods than someone who doesn't have as much amylase in his or her saliva.

Consider your reaction to low-fat foods that are thickened with starch, such as pudding, American-style fat-free yogurt, or low-fat ice cream. It could be that your high level of amylase makes you less enamored of these foods because they just don't give you the same glorious texture transformation you get from higher-fat versions.

Mandel, the lead on the experiment, adds, "On the other hand, people with high amylase do produce significantly more maltose and glucose in the mouth during starch breakdown [meaning their saliva produces more sugar in the mouth when they eat these things], so it's possible that this group finds starchy foods more rewarding and, therefore, likes them more.

"The question of how this perception affects preference and liking for starch-thickened foods is where things get tricky," explained Mandel. "It's likely that an individual is used to his own 'starch viscosity experience.' Generally, people like what they are used to, at least in terms of texture, so it's possible that both high- and low-amylase people equally enjoy the texture of starchy foods, even though the experience will be totally different for each individual." Mandel and Breslin's research gives us the tiniest glimpse into the differing textural worlds we all live in. As we learn more, we'll gain more understanding about how our sense of touch affects the food choices we make.

Irritastes

Gary Beauchamp is the director of the Monell Chemical Senses Center. In the course of his research on taste and smell, he has tasted ibuprofen on purpose, unlike most of us who swallow it whole—and if we taste anything at all, it's the cinnamon coating that brands like Advil apply to the outside of the pill to mask its taste. That's because ibuprofen is very bitter and causes a distinctive sting at the back of the throat. Beauchamp knew this sensory profile well.

One year Beauchamp was attending a molecular gastronomy seminar in Sicily and made a startling discovery about olive oil. Even if you are an aficionado of extra virgin olive oil, you may not have experienced its true essence, which can be captured only minutes, hours, days, or weeks after its virgin pressing. The green blood of olives smells of freshly sprouted flora infused with the newness and glory of life itself. It smells so vibrant that you want to dab it on your wrists, bathe in it, and pour it atop whatever you're eating. There is no scent like it on Earth. I am lucky enough to have friends who have olive trees. I gladly trade hours of my labor picking their olives for one precious, coveted bottle of freshly pressed oil. Roger and I sop it up with bread and sprinkle it on everything in the few days after it's bottled and before it disappears. It quickly loses that vibrancy with age or improper handling. (The worst sin is to store olive oil near a source of heat, where it will not only lose its green top notes but also turn rancid.) Beauchamp had never tasted olive oil this fresh until his trip to Sicily. There, he took a swig of some fresh-pressed oil and was instantly intrigued.

Beauchamp said, "I had considerable experience swallowing and being stung in the throat by ibuprofen from previous studies on its sensory properties. So when I tasted newly pressed olive oil I was startled to notice that the throat sensations were virtually identical."

This throat sensation is a signature of freshly pressed olive oil. It hits you in the back of the throat and lingers well after you swallow. Because it reminded him so much of the sting of ibuprofen, Beauchamp decided to study it further to see if olive oil might have some of the same anti-inflammatory properties as ibuprofen. He did and it does. The low rates of heart disease, stroke, and some cancers in Mediterranean countries may be due in part to ingesting large amounts of olive oil that contains these anti-inflammatory properties.

When something stings, tingles, cools, or burns your tongue, mouth, or throat, you are experiencing chemesthesis. The compounds that provide chemesthesis are literally irritating—not necessarily in a bad way, but they stimulate the same nerve that carries pain information to the brain, the trigeminal nerve. I call them *Irritastes* and they include the spicy-hot heat of jalapeños, ginger, and cinnamon; the sting of carbonation; the cooling from menthol or mint; and, of course, the signature sting of extra virgin olive oil.

Capsaicin is the active ingredient in chile peppers that stimulates the touch nerve. On the one hand, it can be extremely painful, as anyone who has eaten too much sriracha or Cholula hot sauce can attest. In high concentrations it can inflame the soft tissue in the mouth and give a sensation that the mouth is burning. Barry Green, a professor at the Yale University School of Medicine and an expert on tactile and thermal psychophysics, has done a lot of research on capsaicin. When he really pushes his tasters to explain how capsaicin feels, they agree that the sensation is more of a painful ache than a hot temperature. Yet when something burns, we usually associate it with being hot, so we use the same heat-related terminology to describe the tactile burn of chiles. They burn—or savor—hot.

On the other hand, capsaicin in high concentration can be used to treat pain. Various creams on the market, such as Capzasin-P Cream, which claims to provide "temporary relief of muscle and joint pain associated with arthritis, simple backaches, sprains, strains, and bruises," tap into the irritating power of capsaicin. I asked Green how a substance that produced pain could cause relief.

"After repeated exposure, or very high concentration exposure to capsaicin, the mouth or the area that was treated would become less sensitive to it. It was looked at as a potential analgesic [pain reliever] because it was thought it could desensitize pain fibers and therefore render that area less sensitive to painful stimulation of other kinds," he said. "But it's never been proven that capsaicin will penetrate into muscle or into joints to actually desensitize the pain fibers in the muscles or in the joints. We know it can do it in the skin, but it's not clear it can get all the way into the inflamed muscle tissue and actually produce relief." So don't go smearing sriracha over your muscles after a hard workout just yet.

Information about temperature of food in the mouth is also carried by the trigeminal nerve. An inappropriate temperature—a lukewarm cup of coffee or a cold bowl of chicken noodle soup—can completely throw you off. If you'd bought something marketed as Frappucino, cold coffee wouldn't be a surprise. Likewise if supermarkets marketed their produce as "Mealy Macintosh Apples" or "Rock-Hard Honeydew Melons," you wouldn't be disappointed if you bought one and it matched that description.

Unctuousness

It's impossible to fully understand food texture without understanding how fat works. This is only a glimpse, and in the chapter on new contenders for Basic Taste status, I will delve deeper into fat taste, texture, and aroma.

Fat coats the tongue unlike any other substance, which makes the flavors of fatty food stay on your tongue longer as it carries flavor around your mouth. For example, the flavor of strawberries in a fatty ice cream is carried much longer than the flavor of strawberries in a sorbet, which doesn't contain fat. We love foods with fat in part because they savor longer, and this means we can get more flavor out of every bite.

Fat also has a unique melting profile. The way it slowly changes from solid to liquid in your mouth is a type of texture contrast that we're built to love. Fat has nine calories per gram—more than twice as much as the other types of calorie-contributing macronutrients: carbohydrates (four calories per gram) and protein (four calories per gram). Because we are essentially wired to seek out good sources of energy—which is what calories are—we gravitate toward fat. Because fat is so dense with calories it is also demonized by advocates of low-fat diets. An easy way to lower calories in food is to remove fat from it, but that also removes flavor from a food, because fat carries flavor.

Fat is primarily known for transforming food texture. When it's used in baking, it shortens the dough, which means that the proteins in the dough don't develop into long fibers that make it chewy, but instead results in short fibers, which make it crumbly. Think biscuits, scones, and pie crust. Food cooked in hot fat gets its shatteringly crisp texture from the quick heat transfer from the oil into the food. The transfer of heat happens much more slowly when you bake a food. Fat also moves into the air or moisture pockets during frying, which doesn't happen during baking. In fact, some of the fat in or on a food can escape from it during baking or other methods of cooking. Fat makes liquids like coffee creamer more opaque; and because fat can prevent microorganisms from taking up residence, oils and shortenings don't need to be refrigerated.

Many attempts have been made to duplicate fat's unique properties, as I know firsthand because Mattson was on the front line of the war against fat in the nineties. Even our most talented food technologists can do only so much, however. There's just no substitute for the real thing.

Touching

Texture is one of the most underappreciated aspects of food enjoyment. It involves our senses of touch, sight, and hearing. I asked Cameron Fredman, the nontasting, nonsmelling attorney, if he would prefer to add taste or smell if he were miraculously granted one more sense to add to his repertoire of only three. He hesitated a long while before responding.

"I don't know that it would necessarily be an easy choice to get either back at all. What if the world smells bad? What if I didn't like eating anymore if I was given smell?

"I like my life," said a man whose primary experience of food is texture. "I don't feel disabled."

Taste: Identifying Food Without Its Texture

YOU WILL NEED

 4 jars of single-item baby food such as (just) apples, (just) pears,
   (just) peaches, (just) bananas, (just) green beans, (just)
   peas, (just) carrots, (just) chicken, etc.

 Food coloring

 Spoons for everyone tasting

 Saltine crackers and water for cleansing the palate

DIRECTIONS (FOR THE ORGANIZER TO DO BEFORE
THE TASTING)

 1. Remove the labels from the jars and put the product names on
  the bottom so that the tasters cannot see them.

 2. Using the food coloring very sparingly, adjust the color of the
  purees so that they are all very similar shades of brown. The
  objective is to remove all color cues from the food. Go easy
  with the color drops, especially blue and green! It's easiest
  to put a drop of color onto a spoon first, then stir it into the
  food.

TASTE

 • Serve your tasters all 4 of the samples and see if they can
  determine what they are.

 • Have tasters cleanse their palates with crackers and water
  between samples.

DISCUSS

 • Did anyone get all of them right?

 • How did color color the participants' answers?

 • How did removing the texture from a food affect the tasters'
  perception of its flavor?

4

Sight

I'm at dinner and it's pitch-black. Not just dark. Black. My hands are fumbling around my plate. I find something that's clearly an asparagus spear. I lift it to my nose. Yes, definitely asparagus. I continue to move my hands over the plate looking for answers, as if it were a Ouija board. I find something that feels suspiciously like a sponge. Spongy? What kind of food feels spongy? My mind searches for an answer. Morels! Mushrooms. Yes, they are like delicious, hungry sponges, eager to soak up butter sauces. Yes, these squishy things must be mushrooms. But they also might be chicken. It's frighteningly hard to tell. Roger is seated next to me and he *hates* mushrooms, so my fingers walk across the smooth banquet tablecloth to his plate, because I am pretty sure I can find the morels and move them, unobtrusively, over to my plate. If he were to eat one, his HyperTaster reaction to them will not be good. In the pitch-blackness I'm afraid of where he might spit them out.

Then his hand brushes mine. Foiled!

"Why are your hands on my plate?" he asks.

"Um, I think there are mushrooms," I say.

"Really? What do they feel like?"

We are at Dining in the Dark, the name for the annual Foundation Fighting Blindness (FFB) fund-raising event. Proceeds from our tickets would benefit the FFB's efforts to fund research on curing retinal diseases that cause blindness.

Dining in the Dark started in full light. We left our car with the hotel valet as if we were attending any other event. We filed into the banquet lobby and were checked in, given a table number, and handed wineglasses. When we sat down at our table in the main dining room, we exchanged pleasantries with our fellow diners.

Our salads were waiting for us and our sighted server offered us oregano vinaigrette, ladled it onto our butter lettuce with cucumber, and disappeared into the depths of the hotel. We ate our salads in the full sensory spectrum. Then Katie, our server for the rest of the evening, came to our table to replace our sighted salad server. She leaned against the back of Roger's chair and introduced herself. She had been blind since birth. Then the lights slowly went out.

The objective of the foundation's Dining in the Dark event is to put those in the audience into the same sensory situation that their beneficiaries encounter every day. The room went from dark to darker to black so slowly that we barely noticed the change, until we realized we had completely lost our sight. Because the organizers had taped around the doors to thoroughly blacken the room, my chest tightened as I realized that I couldn't leave even if I had wanted to. I was stuck in the dark, suddenly forced to rely on only four of the five senses I use every day.

How does our experience with food differ when we lose one of our senses? This is a really difficult question to answer, because trying to shut off a sense is difficult. There's always the option—and instinct—to remove the blindfold, earplugs, or nose clip. I've learned the hard way that going out in public with your nose or ears plugged will not result in a typical experience. People want to know why you're doing this strange thing. This is what makes Dining in the Dark such a great event. We were all in it together. And there were no blindfolds to remove.

My best friend, Teri, was sitting on the other side of me while the lights were out. She leaned over and whispered to me, "I can't eat anything else."

"Why?" I asked.

"Because I'm completely grossed out," says Teri, who admits to having an extra sensitive disgust-o-meter. She would never eat a sandwich from which someone else had already taken a bite. A hair on her plate will send her out of the room, unable to eat for the next hour without retching.

"What?" I ask. Something on her plate must be fueling the gross-out.

"I don't want to be insensitive or sound like I'm trying to be ironic, but there's something on my plate that feels like a human eyeball."

I poked around my plate and found that, indeed, she was right. The chef's choice of a roasted cherry tomato as garnish for the chicken was probably not intended to have this ghoulish effect, but it did.

Since I began working in the field of food development, I've known that sight is critical to eating. In any food prototype-tasting meeting, someone will almost always say, "People eat with their eyes." Usually, this is in response to a prototype that looks like crap, sometimes literally (chewy chocolate candy logs, you get the picture). Yet when we're talking about visuals, we're usually talking about how the color of a food or package or plate will affect how much consumers like it. By dining in the dark, I learned that it's unbelievably difficult to discern what you're eating without your eyes.

Is it really that hard to tell without your eyes that a mushroom is a mushroom and a carrot, a carrot? If you were to lose your sight suddenly, the answer most definitely would be yes. When the lights went out in the hotel banquet room, the sighted people became helpless. But most people are not thrown so jarringly into blindness. For example, people with retinitis pigmentosa, one of the diseases that FFB is trying to cure, lose their sight gradually. The gradual loss enables them to learn how to compensate with their other senses.

Janni Lehrer-Stein chaired our dinner. A five-foot-tall bundle of energy with long, curly brown hair, she describes herself as a foodie and a Jewish mother, two things I know and love firsthand. She's whip-smart, funny, and does good work, serving on more than a few nonprofit boards, the Foundation Fighting Blindness being just one of them. I liked her the instant I met her.

Janni was twenty-six and practicing law in Washington, D.C., when a snowstorm that she will remember forever hit town. Unable to get out to a doctor to treat an eye infection, she went to see the ophthalmologist who lived in her apartment building. He dilated her eyes and pronounced very definitively, "You will be blind in six months."

Luckily for Janni, he was wrong about how long she would remain sighted. But he was correct about the disease she has that will steal her last bit of sight any day now. Janni is in her fifties, living very comfortably in San Francisco with her husband and three children. If she bumped into you in the grocery store, she would look you straight in the eye and apologize in her warm and friendly manner, and you would probably just think she was klutzy, not suffering from a blinding disease. Janni still has "one last lap" in her, which is how she refers to the time left before she will be completely blind. She can see shapes, see contrasts, and read print if it is big enough and she can get close enough. But all of this is deteriorating.

To learn how Janni chooses her foods and cooks, I called and asked her if she would cook with me.

When we went grocery shopping, Janni parked her big black guide dog, Nanaimo, outside Cal-Mart and grabbed a cart with the confidence of someone who has been shopping at the same store for decades. Janni doesn't shop there because it's the closest supermarket to her home, but because they don't move things around frequently, as other stores tend to do. Once, a few years ago, after the produce section had been rearranged, she was forced—overnight—to relearn the geography of the fruits and vegetables from scratch.

As we shopped in the produce section, Janni picked up plastic fruit containers, opened them, and fingered the individual berries before confirming that they were the correct ones she'd use to make a raspberry chicken entree. She grabbed bunches of beets while telling me the story of a beautiful salad she made recently. She had envisioned the earthiness and gorgeous red color of cooked beets, paired with contrasting sweet and sour orange segments, but when the salad was complete, she knew something was wrong. Her children laughed when she asked them why the red beets and the oranges didn't look right. "Because they're yellow beets," they told her.

Janni and I continue to the tomato section and I watch her pick up tomatoes on the vine, then put them down and choose a firm heirloom instead. Janni didn't fall for the smell of the vine that fools consumers into thinking the tomato has more volatile aroma than it really does. I wonder how much of the produce I buy is driven by how pretty it looks. If farmers were forced to market their wares specifically for the blind, we sighted people would also be the beneficiaries of tomatoes that would most definitely savor better.

Dictator of the Senses

Our sense of sight confounds what we think we savor. In test after test, researchers have proved that we default to our visual system even when our taste and smell systems work just fine. When consumers taste miscolored beverages, they have trouble identifying the flavor. If you are given an orange-colored glass of apple juice to savor, you are likely to say that that you taste orange juice. If you are given a glass of clear but flavored liquid, you're unlikely to recognize it as cola even if the beverage tastes exactly like the cola you drink every day. With other foods besides beverages, such as jelly, sherbet, and candy, each study comes up with the same results: changing the color of a food or beverage can change what you savor.

All these tests were done with normal consumers. You might think that professional tasters would not fall prey to this phenomenon. But they (we!) do. In two separate studies, both beer and wine professionals were completely fooled by their own eyes.

In one study, participants were given nine sample cups of three brands of beer. There were three cups of Pelforth (French), three cups of Ch'ti (French), and three cups of Leffe (Belgian). The researchers added flavorless color to the cups so the result was nine cups of beer in three color groupings: three light-colored, three medium-colored, and three dark.

First the beer professionals were asked to taste all nine beers without being able to see the color. The tasting session was done in a booth that was lit with colored neon, which rendered the beer color imperceptible. And just for good measure, the beers were served in black cups that further obscured their color. This way, the tasters were not clued in to the researchers' attempts to hide some visual aspect of the product, as blindfolding the tasters would have done. Next, the participants were asked to taste the beer in a more normal situation, where they could see the color. In both cases, they were asked to sort the beer into category groupings.

When participants were able to see the beers, they sorted them by color.

Even more interesting, when asked why they sorted them into the three different color groups, they didn't report color as a reason. They thought they were using their professional sensory evaluation skills to separate the beers into what were clearly, for them, three flavor groupings.

When they tasted the samples without the benefit of seeing them, though, they sorted them by brand: in other words, by flavor. Without their eyes to function as the Dictator of the Senses, they had the freedom to use their senses of smell and taste. Even professional tasters' sense of sight can overrule the actual sensory input they get from food.

Another famous color study was conducted on wine at l'Université de Bordeaux in France. The researchers came up with a lexicon of descriptive terms for white wine (lemon, lychee, butter, white peach, and citrus) and for red (prune, clove, cherry, chocolate, and tobacco).

The researchers chose one characteristic white wine, an AOC Bordeaux vintage 1996 blend of semillon and sauvignon blanc grapes, and a characteristic red Bordeaux wine (also 1996) that was a blend of cabernet sauvignon and merlot grapes. Then they added flavorless red color to the white wine to match the color of the red wine to create a third entrant for the taste test. This third wine, the red-colored white, was different from the white wine only in color.

The wine tasters were given the list of wine terms and given samples of each wine in pairs. They were asked to choose which wine most closely fitted the descriptors. The tasters perceived the red-colored white wine as having the flavor of a red wine. The authors write, "The wine's color appears to provide significant sensory information, which misleads the subjects' ability to judge flavor."

Why are we so easily misled by our eyes? One reason is that our sense of sight is simply quicker than our sense of smell. Our ability to detect something visually—*that it is a glass of wine and it's red*—happens ten times faster than our ability to detect an odor. We humans are spectacularly bad at identifying smells without any other sensory clues, as demonstrated by the *Top Chef* Blindfold Challenge and my Spice Rack Aroma Challenge. We're also really bad at describing odors even when we recognize them and they trigger strong emotional reactions. Saying, "It smells like Uncle George" or "It reminds me of third grade" is charming but doesn't really give concrete information about what the odor is. We're just not very good at applying our incredibly acute sense of smell to taking rudimentary tests. In contrast, it would be rare that you'd look at a glass of red wine and erroneously say it looked white. We're very good at visual identification.

The senses of sight and hearing function at a distance. You can be seated dozens of feet from the stage of *Iron Chef*'s Kitchen Stadium and still get the full benefit of the experience visually and sonically. Taste doesn't work this way. Even if you can see and hear the chefs chopping onions, you can't taste onions from a distance. To taste them, you have to put them in your mouth. Taste, like touch, is a contact sense. In order for the sense to be activated, something has to come in contact with a receptor.

Smell works both ways. You can smell onions at a distance as well as at close range, but the distance at which you can smell onions is far shorter than the distance at which you can see them. Once again, sight trumps smell.

We naturally rely more heavily on sight than smell or taste because of a distinction between the external and the internal. Sight and hearing happen outside us, but we need to put food close enough to smell or inside our mouths to taste and smell it. And this usually happens after we've seen it; that is why it's so hard to override your eyes. When you see an orange beverage, you expect it to taste orange. When the beverage doesn't match that expectation, you get confused. People who are asked to taste appropriate-colored drinks (orange-colored/orange-flavored) responded faster than those asked to taste inappropriately colored drinks (green-colored/orange-flavored), proving that when we can't default to vision, our brain searches for more information—not always an easy task when the necessary input is our keen but emotion-fraught sense of smell.

Sensory Snack

This was my favorite quote from all my scientific research:

There was no neutral food because food, regardless of what it is, is not associated with neutrality but always with some emotional intensity.

I. C. U. 8.

My colleague and fellow foodie, Candice Lin, brought to work a can of silk-worm pupae, the cocoon stage of the silk moth's growth, which comes after it is a caterpillar and before it becomes a flying adult silk moth. Lin had been shopping at a Korean grocery store and found the product so appalling that she simply had to buy it. And for some reason, she felt compelled to open the can of critters only when I agreed to eat one with her. I had been desperately seeking surströmming, so a few insect cocoons weren't going to scare me off. After all, the package said they were "High Protein—Great side dish when drinking alcohol," making them sound a lot like salted peanuts.

Another colleague, Rich Gorski, overheard us talking and before either one of us girly girls had worked up the courage to eat a pupa, reached into the can, grabbed a bug from the murky liquid, and popped it into his mouth. A consummate professional, he chewed, he chewed, and he chewed some more, the look on his face absolutely neutral. Not good but definitely not bad. Lin and I reached in, buoyed by his apparent acceptance. Lin beat me to the chew by about twenty seconds, and her immediate reaction was definitely not neutral. A look of utter disgust crossed her face and I knew the minute I bit into mine—while watching her chew, grimace, and flinch—that this was not going to be good. I sank my teeth into the pupa, gave a few chews, and ran to the sink to spit it out.

Upon reflection, this little mummified worm was no worse in flavor than the crispy fried grasshoppers called *chapulines* I had eaten in Monterrey, Mexico. In fact, with a bit of lime juice and dried chile powder they might have tasted quite similar. Why was I so influenced by Lin's pained face but not by Gorski's neutral one?

Researchers in France set out to see if the emotions on someone else's face could influence the desire to eat a particular food. They chose three foods that their panelists liked (bread, green beans, chocolate bars) and three foods that the panelists didn't like (bloody red meat, kidneys, blood pudding). Then they paired each of these foods with one of three human faces: one showing pleasure, one showing neutrality, and one showing disgust. The emotional faces proved to be very powerful. In the presence of disgusted faces, the test subjects wanted to eat the liked foods *less* than they did without the emotional face present. They wanted to eat disliked foods *more* in the presence of pleasure.

You could argue that emotions displayed on another's face gives the "reader" information about a food, but this study was conducted with six foods that were everyday, familiar foods to the participants, who already knew how they savored. And no one was actually eating the food, so you could assume that the participants weren't worried about its safety or freshness. I think there is another communication-based explanation for this.

The *embodiment theory* says that seeing an emotion on another person's face will induce that emotion in the observer. In other words, when you watch someone in pain, you experience a bit of pain yourself. If you were to apply this to eating, you would expect that watching someone struggle to keep from retching after eating a silkworm pupa would make you a bit nauseated as well. The same would hold true if you saw pleasure on the face of someone who was eating grasshoppers, as I did in Mexico when I was eating with a Mexican national, Giselle Marce, a client who had grown up with chapulines. We were also drinking. The grasshoppers are served as a salty snack with alcohol, which seems to be insects' natural habitat on a menu. Marce's nonchalant ingestion of them made it easier for me to get them down than the pupae.

There was a similar study that coined the term *chameleon effect*. The researchers got participants to mimic their physical behavior—rubbing their nose or shaking their foot—simply by doing the desired behavior in the course of an unrelated task. If I sit across from you and scratch my nose, you are likely to do so, too. If I tap my fingers, you might tap yours as well. If you doubt that this is a basic human behavior, the next time you're with a friend or in a meeting at work, put your hands behind your head, spread your elbows out, and lean back. It's almost comical how quickly the people around you will do the same. This is a result of unconscious mimicry, which the study authors related to the ability of a chameleon to blend into its surroundings. To quote from the study: "The chameleon effect serves the basic human need to belong . . . a powerful, fundamental, and extremely pervasive motivation."

We human beings are incredibly good at detecting emotion on others' faces, we wince at others' pain, and we react as if we are feeling the same thing as someone else. But we aren't really in touch with how much these emotions can influence our behavior. We use our eyes to read emotions on the faces of others as we eat. And the minute we see something, we react whether we're conscious of reacting or not.

This is relevant if you are raising children and want them to be open-minded about what they eat. Children want to belong and they're prone to mimicry from an early age. As a child I watched my mother make pained faces after eating peas or green beans and so wouldn't eat them until I was in my early twenties. I didn't know they could taste delicious. You, too, can unconsciously telegraph your food likes and dislikes to your family. This can be useful if you work it consciously, for good, not evil. By expressing obvious joy and delight to your family when eating Brussels sprouts and salmon you can help increase their desire to eat these healthful foods. If you want your kids to avoid a certain food, eat it with them and visually communicate your disgust. In the same way that my mother telegraphed to me that peas were bad, you can communicate to your kids that goo-goo cakes are bad.

In the professional field of sensory evaluation and product development, we know this chameleon effect all too well and have to adjust for it. When we conduct taste tests on food prototypes at Mattson, we set up independent booths for the testers to sit in so they can't see one another's faces. We know all too well that one "eww" face can influence a whole roomful of people, so we try to prevent this from happening.

Single-Sense Shopping

For sighted people, grocery shopping is mostly a visual exercise. Because you can't sample the tomatoes, milk, or bread, you simply have to use your eyes to ensure that what you're buying is what you want. In the produce section, we can use our sense of smell with some fruits and vegetables (peaches, tomatoes), but not all (onions, carrots, bagged lettuce). For packaged foods, we rely on the photographs or images on the front to tell us what's inside.

This single-sense shopping experience makes our grocery stores complicit in encouraging us to eat the same stuff we've always been eating. There simply isn't enough over-the-counter and in-aisle sampling to nudge us out of our comfort zone and into the experimentation mind-set necessary to buy a bag of Jerusalem artichokes or a head of black garlic or piece of skate wing.

How in the world would you know how they savor in today's sterile super-markets? Some retailers do a better job than others, offering samples in the aisles. But I have a vision of a future where the shopper can sample every single thing in the store before buying. If real, edible samples aren't feasible, I'd even settle for scratch-and-sniff. At least we'd be one sense closer to eating. The supermarkets of today will be looked back upon as bland, boring, sensory-deprived morgues. Whole Foods has a loyal fan base due in part to the fact that it's turned grocery shopping into much more of a sensory experience than its traditional competitors. But even Whole Foods hasn't gone far enough, in my opinion. Eventually retailers will embrace both low and high technology to approximate the experience of eating using multiple sensory inputs at the point of sale, in contrast to today's stores, where we rely on sight almost exclusively.

The See Food Diet

You decide how much to eat using your eyes. Brian Wansink, director of the Food and Brand Lab at Cornell University, does research on how people decide what to eat. One study on portion size used a deviously brilliant self-refilling soup bowl. Imagine a table with a floor-length tablecloth over it. The participant sees only what looks like a normal bowl of soup on the table, not the mechanism under the table that continuously refills the bowl as it empties.

Wansink measured the amount of soup that test subjects consumed when eating out of a normal bowl and out of one that never emptied out. The result was not surprising, but the magnitude of the difference was: participants ate 73 percent more soup when the bowl stayed fairly full. When asked afterward to estimate how much they'd eaten, they estimated that they'd eaten the same amount as those eating from normal bowls. Worse yet, they didn't notice being more full when they'd consumed the extra calories. The results were crystal clear: "People use their eyes to determine how much they eat. This biases their intake and can lead to overconsumption." How much of the obesity problem is due to that old adage, "Your eyes are bigger than your stomach"?

We get information about what a normal portion size is from the visual appearance of a food. Our eyes tell us that a full plate of food from a casual dining restaurant is a normal meal. We adjust our eating to that norm, even if that norm may be enough calories to sustain us for an entire day. We look at a whole candy bar as a normal treat, even though that candy bar might be three times the size of candy bars from twenty years ago. We have yet to connect our behavior to this simple fact: bigger portions, bigger packages, and bigger plates result in bigger waistlines.

The Cornell researchers have a suggestion on how to avoid these visual traps. They can trick people into eating 73 percent more with a bottomless bowl, so it's likely you can trick yourself into eating less by using smaller plates, bowls, and glasses. Repack snacks (or other bulk foods) into smaller baggies that set the appropriate portion for you to eat. In fact, the food industry itself took heed of this insight and introduced single-portion packages in the mid-2000s. You can see 100-calorie packages today on the grocery store shelves. The researchers also suggest that leaving empty bottles of wine on the table will remind guests that they've had enough to drink. The reverse should also hold true for restaurants. If you want your customers to order more beverages, clear the empty glasses and bottles from the table as quickly as possible. It seems that we're genetically programmed to react to an empty glass or plate the way nature abhors a vacuum.

Cooking Without Looking

Janni, the chairwoman of FFB, and her family live in Pacific Heights, one of San Francisco's grandest neighborhoods. Their home is magnificent, four stories tall, with a kitchen that would make a sighted foodie weep with envy. Upon closer inspection, this is a kitchen for a woman who will soon be completely blind. The white marble countertops were chosen for their blank sheet-of-paper functionality. Janni placed a pea on each piece of stone she was considering before deciding that white was the best color for finding small food objects with her remaining limited vision. The counters are beveled and smoothed around the corners so she doesn't bruise her hips when she bumps into them.

Cooking with Janni is a contact sport. When you can't see what you're doing in the kitchen, the best utensil to use is your hand, which gives you more input from the sense of touch. She mixes mashed potatoes and sour cream together for the knish filling we're making in a way that reminds me of kids playing with wet sand on the beach. She considers herself a baker and proudly shows me the enormous fifty-pound bags of flour and salt that she keeps in drawers for easy access. The sugar shares a drawer with Nanaimo's kibble, which immediately conjured up for me a visual image of a pitcher of "sweetened" iced tea mistakenly made with tea, lemon juice, water, and dog food.

The few things Janni doesn't do with her hands involve sharp objects. The safety lid of the Cuisinart food processor takes on new meaning when you're blind. Janni's knife skills are charmingly lacking, I notice as I stir the onions she's chopped up, removing the tufted end of the onion and a few bits of foil butter wrapper from the sizzling pan. As we fill the knish rounds, I notice that hers are bigger but better sealed than mine. She's more experienced than I. Even without her vision, she's learned how to cook without looking.

There's still controversy about whether blind people develop more acute senses of touch, smell, taste, or hearing than sighted people; regardless, they do use the four remaining senses more effectively.

And hopefully, a healthy understanding of how sight influences what you think you're savoring—and how much—can help you use all your senses more effectively.

Taste: Can Color Color Taste?

The goal of this exercise is to experience how changing the color of a juice can confound what we savor. Tell your tasters it's okay if they get the answers wrong. That's part of the fun.

In preparing for this exercise, you will be "making" 5 cups of juice in 5 different colors. They all need to look natural, like various types of juices. Go easy with the color drops! It's best to put a drop of color onto a spoon first, then stir it into the juice. That way, you won't squeeze out too many drops by mistake. Be sure to prepare the juice—and hide the bottles—before your tasters arrive.

YOU WILL NEED
    Masking tape and markers
    5 transparent cups, glasses, or bowls
    ½ cup apple juice
    ½ cup lemonade (which contains only lemon juice, water, and
        sugar)
    ½ cup white grape juice
    ½ cup cranberry juice cocktail
    ½ cup pear juice (or other light-colored juice)
    1 package food coloring (containing blue, red, yellow, and
        green)
    Spoons for tasting
    Saltine crackers for cleansing your palate
    Paper and pens or pencils

DIRECTIONS
1. With masking tape on the bottom of the glasses, mark each juice by name.
2. Add ¼ cup of the corresponding juice to each glass.
3. To the glass of apple juice, add 2 drops of red color (goal: red).
4. To the glass of lemonade, add ½ drop of red color (goal: pink).
5. To the glass of white grape juice add 1 drop of red, 1 drop of yellow, and 1 drop of blue (goal: brown).
6. To the glass of cranberry juice add 1 drop of red and 1 drop of blue (goal: deep purple).
7. To the glass of pear juice add 1 drop of yellow (goal: orange).
8. Tell the tasters you'd like them to tell you what type of fruit juice they taste. Tell them to be as specific as possible. For

example, instead of just orange juice, you want to know what type of orange it is. Navel? Tangerine? Have them cleanse their palates with saltines between tastes.

9. After the tasters have written down their guesses, let them know what is what.

OBSERVE AND DISCUSS
- Did anyone get all of them right?
- How did color color the participants' answers?

Taste: The Blindfolded Diner

YOU WILL NEED
1 willing cook or meal procurer
Whatever meal you choose to cook or serve
1 willing participant
1 blindfold

EXPERIENCE, TOGETHER
1. Cook a meal for someone (or a group of people) you love. Try to contain the smells away from the area where you'll be eating so as not to tip off what you're cooking.
2. Sit down in the dining room with your willing guinea pig(s) and explain what you'll be doing. It's really important that you have their trust and that you do not betray it. Tell them that everything you're going to feed them is safe, familiar, and hopefully delicious. Then, make it so!

This is not the time to introduce to your loved ones a new food, or try to get them to eat something they hate. That's not the object of this exercise. The object of "eating blind" is to see if you can recognize familiar foods without your eyes.

1. While you're sitting down with your diner(s), have them get familiar with the table, place setting, and position of the glasses. Don't blindfold them until they're 100 percent comfortable with their surroundings. Have them stay in that seat for the duration of the experiment.
2. Serve your diner(s) as you would do at a restaurant, one course at a time. Sit with them and watch them eat. Be there to answer any questions. After they've tried to guess what they're eating, you can tell them what it is.
3. Have them eat it again—still blind but knowing what they're eating—before you remove the blindfold.
4. Then remove it and let them enjoy the rest of the course with their sight. Blindfold them again before the next course arrives.
5. Enjoy. Have fun with it.

5

Sound

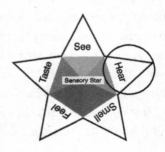

Don't Sell the Steak—Sell the Sizzle

Über salesman and author Elmer Wheeler, 1937

How do you choose a restaurant? Maybe you visit a site like Zagat or Open-Table. Or perhaps you type into your search page a city name plus a trusty phrase such as "best restaurant in" or "top 10 restaurants." Regardless of where you end up on the web, you're likely to have a fairly limited sensory experience, which makes your choice difficult because dining in a restaurant is one of the most complete multisensory experiences available to us on a daily basis. Until technology advances, you won't be able to smell or taste the food from the restaurant in question. You may be able to tour the restaurant using your eyes if the owner has the foresight to post photos or, even more rarely, a video. Some restaurateurs will even include a signature song or soundtrack that puts you in the mood. But that's likely to be the extent of the sensory input you get to help you make your choice.

Delfina restaurant has approached the online medium differently. With the neighborhood's signature sounds, the Delfina website transports you to the scrappy, bustling corner of Eighteenth Street and Guerrero in San Francisco's edgy Mission District. You feel the energy. You sense the pulse of the place. You hear the sound of chairs scraping across the hardwood floor, forks clanging on plates, place settings being cleared, glasses clinking, phones ringing, and the constant din of a happy crowd of pasta-twirling, wine-drinking customers. The noise level is just right: energetic but not too deafening, infectious in a way that makes you want to go there just to be part of the action.

But Delfina didn't always sound this way. When it first opened in 1998, San Francisco restaurant critic Paul Reidinger described it as "spare walls, stone floors, and shiny, cold zinc-topped tables amounting to an ideal environment for the propagation of decibels . . . a crescendo that's not unlike the approach of a train. All that's missing is a horn and a flashing light." Owner Craig Stoll remembers another critic calling Delfina a "shrieking hellhole."

It is a testament to the quality of Stoll's food that the place survived at all, because eating there was a painful experience. The restaurant had the same throbbing, audible excitement as it has today, but at its original size—about one-third as big as it is today—the noise was stuffed in, uncomfortable. Today, Stoll gestures proudly at Delfina's ceiling, bar, and walls.

"You're looking at ten thousand dollars' worth of acoustic panels," he says. "This place is loud," he continues, "but the sound is good. It's a good-quality sound."

Even with the moderated sound, Stoll had to engineer his menu to compete with the space's energy. "There's sensory overload in this place. And we like it that way. There's a lot going on. There's a buzz around you, and this plate of food has to compete for your attention. Our food's pretty subtle, but it's just past subtle."

Stoll is onto something, according to a study conducted by the food company Unilever and the University of Manchester. They wanted to find out whether background sounds affect the perception of flavor. They found that people rated foods less salty and less sweet as noise levels increased. When noise levels decreased, the perception of those tastes increased. The results indicate that noise has a somewhat masking effect on taste. This is one of the reasons why airplane food doesn't taste very good. The deafening roar of the engines can make the food taste less sweet and less salty (and possibly less other stuff, too, that these researchers didn't test for).

There are other reasons, of course, why food at 30,000 feet isn't very good. There's much less humidity, which dehydrates you. This translates to less saliva available to dissolve Basic Tastes and less mucus in your nose to absorb aromas. Most of the hot food was prepared and cooked hours before you will eat it. But certainly the droning of two jet engines plays a part in overpowering subtle tastes. Similarly, when Stoll pushes his food "just past subtle," he's turning up the volume of his food to compete against the masking sounds of his restaurant.

A company called Acoustiblok sells something they call QuietFiber for combating this type of flavor-masking cacophony. They recommend cutting QuietFiber into strips and squares and securing it all around a loud restaurant: under tables, under the bar, under chairs, and so on. The next time you feel the underside of a table at a restaurant, what you'll end up finding is, hopefully, the result of someone thoughtful like Craig Stoll. Finding noise-absorbing panels in a restaurant is a good sign. It means the proprietor is thinking about more than just the two or three senses you primarily use when eating. He knows that a mellower decibel level allows you to experience flavor more fully. He also probably knows how to use each human sense, balancing them as carefully as he does the five Basic Tastes that you'll savor in his food.

Sensory Snack

Lufthansa tests their airplane food in a grounded plane that simulates the atmospheric pressure, sound, temperature, and humidity present at 30,000 feet.

The Sound of Music

If your sense of taste is offended, you can spit out an unappealing food. You can pinch your nose when an awful odor overwhelms you. To shut out offensive images, you can simply close your eyes. But since you have no earlids, your sense of hearing is often assaulted without your permission. Julian Treasure, founder of The Sound Agency, has worked with food retailers such as Tesco and Marks & Spencer to help them reengineer their spaces to reduce negative sounds and enhance pleasant ones. "There's no point in playing a pleasing soundscape or some appropriate and effective music on top of a cacophony of other noise. That's what I call 'putting icing on mud,'" says Treasure.

Treasure tells a story about friends of his who walked into a very quiet hotel bar in London. As soon as the party of four sat down, the bartender turned on the music—a pumping dance track—at a level so loud they ended up shouting at each other. When they asked the bartender to turn the music down, they also asked why he had turned it on in the first place. They were the only ones present and they hadn't asked for it. It turned out that management had instructed all employees to turn the music on as soon as customers walked into the bar "to create atmosphere."

"That is the most misused word in the restaurant business," says Treasure. "It's often used synonymously with noise. Loud music is not equal to atmosphere." What Treasure means is that music—in the absence of activity, energy, and, most important, customers—does not instantly bring a bar or restaurant to life. If only it were that simple!

Loud music may make the environment less pleasant to some people, but it can positively affect sales of alcohol. In a study conducted in two different bars, the researchers found that revelers ordered more drinks and drank their beer faster when the music playing in the background was fast and loud. When the sound track was played at a lower decibel level, drink sales were lower and the pace of drinking was slower. In other words, fast tempos beget fast-moving partiers who also, not incidentally, spend more money on drinks.

Musical tempo also has an effect on the pace at which diners eat food. So, if restaurateurs want their customers to linger longer, they should play slow music. Conversely, if their objective is to get you in, get you out, and turn over your table, playing fast music will help. Next time you're assaulted by frenzied club music in a crowded restaurant, you'll know you're being given a not-so-subtle hint to eat and skedaddle.

The takeaways from these studies refer to averages, however, and it's dangerous to assume that one or two people will respond in an average way. Treasure cautions:

> The effect of sound will vary hugely depending on occupancy. For example, the impact of switching on loud music where there was none in an almost-empty bar is going to be very different in a busy bar with that music and a busy bar with no music. Where the customers experience the change in condition, that itself creates an effect—in the case of my friends, an adverse one. They were more minded to leave than to drink faster!

He also notes that other factors are at play. Do the customers like the music? Is it appropriate? For example, would Queen's rock anthem "We Will Rock You" add the same thrilling vibrancy to fine dining as it does to sports arenas? Probably not. The quality of the sound is important, too. No matter how perfectly the music is matched to the food, a tinny sound says to your customers that you don't believe in delivering a good-quality sonic atmosphere. When I can see, hear, smell, taste, or touch something on which a restaurant has clearly economized, I wonder in what other areas it is skimping. That's a train of thought you don't want restaurant customers taking.

My favorite study on how sound affects flavor perception was funded by Aurelio Montes, of Montes Wines in Chile, a man who believes so strongly in the power of music that he bathes his aging casks of premium cabernet sauvignon wine in the sound of Gregorian chant. Mr. Montes inspired a professor to test whether or not music could influence the taste of wine. The findings could be a boon to aging hair bands across the world. It turns out that listening to a heavy metal song while drinking a cabernet sauvignon—Guns N' Roses's "Sweet Child O' Mine," for example—can make the wine taste more robust. According to one theory, the wailing sound of Axl Rose lights up certain areas of your brain that, for example, might correspond to heavy, hearty, robust, and muscular. This stimulation then primes your brain to taste wine in the same way.

Wine Varietal	Suggested Songs for Optimizing the Experience of the Wine
Cabernet sauvignon	All Along the Watchtower (Jimi Hendrix)
	Honky Tonk Woman (Rolling Stones)
	Live and Let Die (Paul McCartney and Wings)
	Won't Get Fooled Again (The Who)
Chardonnay	Atomic (Blondie)
	Rock DJ (Robbie Williams)
	What's Love Got to Do with It (Tina Turner)
	Spinning Around (Kylie Minogue)
Wine Varietal	Suggested Songs for Optimizing the Experience of the Wine
Syrah	Nessun Dorma (Puccini)
	Orinoco Flow (Enya)
	Chariots of Fire (Vangelis)
	Canon (Johann Pachelbel)
Merlot	Sitting on the Dock of the Bay (Otis Redding)
	Easy (Lionel Ritchie)
	Over the Rainbow (Eva Cassidy)
	Heartbeats (Jose Gonzalez)

Source: Montes Wines

This type of research on sound has such delicious implications that chefs are already putting it into practice in the field. One such chef is Heston Blumenthal of The Fat Duck in England, which in 2010 took the number three spot in San Pellegrino's survey "The World's 50 Best Restaurants."

Blumenthal has worked closely with Charles Spence, the professor who heads the Crossmodal Research Laboratory at the University of Oxford. *Crossmodal* refers to how one of our sensory modes, such as sound, can cross sensory lines and influence another, such as taste. Together, Blumenthal and Spence crafted two experiments to illustrate how environmental sounds can influence flavor perception. The first one used a flavor of ice cream normally not found in nature: bacon and egg. Chef Blumenthal's ice cream is served at

The Fat Duck with a piece of fried bread, which unifies the dish and adds a crispy textural component that conjures up actual bacon and eggs. In the experiment, conducted at a conference on art and the senses, participants were asked to rate the "egginess" and "bacony-ness" of the ice cream while one of two sound tracks played in the background. When the sound of bacon sizzling, popping, and crackling in a pan was played, the tasters rated the bacon flavor higher than the egg flavor. When the researchers played a sound track of barnyard chickens clucking, eaters rated the egg flavor higher than the bacon. It seems that you can pull a flavor in one direction or another with auditory bait.

The second experiment was geared toward determining if they could manipulate the pleasantness rating of a food that, in the absence of culinary accoutrements, can look horrifyingly unpleasant: a raw oyster. Here I will invoke Jonathan Swift, who claimed that the bravest man who ever lived was the first one to eat a raw oyster. Who in his right mind would have thought to crack open what looks like a prehistoric rock formation, only to find inside it a quivering, gelatinous, slimy gray mass—and think to himself, *That looks delicious?*

The first oyster in the experiment was served on the half shell, the way most restaurants serve them. In the background a sound track of waves crashing on the beach was playing. The second oyster was served in a petri dish, making that quivering, gelatinous, slimy gray mass look like an organ being readied for transplant. In the background they played the discordant sounds of clucking chickens. Not surprisingly, the participants rated the oyster on the half shell with ocean sounds much more pleasant than the petri dish oyster with clucking sounds.

Blumenthal demonstrates the oyster test results daily at The Fat Duck, when he serves a seafood course called Sound of the Sea. Diners are presented with a large seashell inside which is an iPod. Then they're served a glass dish of edible foam and fresh seafood perched atop a box of sand. Diners are instructed to put on the iPod earphones to hear the sounds of the sea before digging in.

Blumenthal and Spence note that the dish does three things. First, it makes diners think more about the effect that sound has on the appreciation of food, something we often take for granted.

Second, as proved in their research, the soundtrack intensifies the seafood-y flavors in the dish. The sound of the waves lapping the beach transports you to the seaside, conjuring up aromas of salt spray and ocean air, which you ascribe to the food you're eating.

And third, the earphones make diners focus on the dish more than on companions or conversation.

Eating Loudly

Your sense of hearing is also important once you put food in your mouth. Imagine eating a potato chip without hearing the shattering crispiness, a bowl of cereal without the cacophony of flakes and clusters collapsing in the mouth, or a bag of pretzels without the head-throbbing crunch. This is one of the reasons that popcorn is the snack of choice in movie theaters. Imagine trying to eat a bag of Doritos in a theater. You'd probably get some angry glares. In fact, most theaters don't even sell crunchy snacks at the concession stand. They're simply too loud, both inside your head (affecting your ability to hear the movie) and outside your head (affect your seatmates' hearing). Popcorn, on the other hand, compresses demurely in your mouth with a quiet, unobtrusive crunch that doesn't interfere with the movie sound track.

> ### Sensory Snack
> The human ear is so sensitive that people can tell the difference between hot and cold coffee simply by listening to the two being poured.

As annoying as loud eating can be, the sounds of people eating can communicate a lot of information about their food. In laboratory studies, people who simply listened to the recorded sound of someone eating celery, turnips, and crackers gave the foods the same texture ratings as those who actually ate them. You can use this knowledge the next time someone is smacking his food obnoxiously loudly by saying something along the lines of, "I can tell from the loud sound of your chewing that the chips you're eating are really fresh." That might shush him down a bit.

If you wanted to conduct a study on how sound influences the perception of potato chips, you'd have to standardize the stimulus: the chip. If one tester got a thick, folded chip, his experience would be very different from that of another tester who got a thin, flat chip. Here—again because of their uniformity!— scientists have come up with the perfect solution: Pringles. Because each double saddle–shaped crisp is identical, Pringles are a food researcher's dream. One study showed that consumers rated Pringles crisper and fresher when they heard loud sounds of them being eaten. Chips with lower sounds were more likely to be rated as stale or soft. The same test was done with carbonated water. The louder the sound the bubbles made, the fizzier the water was rated.

This research led me to some obvious questions: What happens to a person's experience with food if his hearing is compromised? Does he experience crispy foods as less crispy, bubbly foods as less bubbly, and so on? I couldn't find any research on this, so I decided to do my own.

The Association for Late-Deafened Adults (ALDA) is composed of just the people I was seeking: those who had been born hearing but had lost this sense later in life. When I went to their San Jose, California, meeting, I discovered that losing their hearing had affected their experiences with food, but not in the ways I had anticipated.

"I can't hear the teakettle whistling," said Jim Letter, a former bartender, "so I have to watch for the steam."

"There's things—like the watermelon that we used to bang on to hear if it has that right hollow sound—that we can't do anymore. So we have to assume that it's going to be okay," said Linda Drattell.

Could they still hear the crunch of a potato chip after losing their hearing?

"You *feel* it crunch in your head," said Drattell. Tiffany Freymiller told us that she seeks out a Pop Rocks–like topping at the yogurt shop because she likes the way the candies feel as they're popping. "I like the way they pop. I can't hear them, but they're adding something."

Everyone in the group agreed that restaurants play a large role in their social life as late-deafened adults. Until recently, most movie theaters, plays, and concerts could not accommodate deaf people. Dining is one social event they can fully participate in, though it has frustrations, one of which is the lack of written specials.

"There are no gold stars that waiters get for memorizing the list and spitting it out quickly," said Drattell, annoyed at having to read the lips of servers speedily reciting the specials of the evening so that they can get on to the next table.

The minimalist-decor trend and open kitchens generate a lot of noise that also poses a challenge. The hearing-impaired adults I spoke to wished for a quiet place set aside where they could dine in relative comfort. Yes, people who can't hear want quieter restaurants. The bouncing cacophony of a bustling restaurant overwhelms what little hearing they've got left and completely frustrates hearing aids. Noise creates the illusion of energy for people with normal hearing, but frustration for those without.

The population that's deaf is fairly small, but the population that's aging and suffering from hearing loss is skyrocketing. A 2010 study from the University of Wisconsin reports that 37 percent of men born between 1944 and 1949 suffer from a hearing impairment. These researchers also predict that 50.9 million Americans will be hearing impaired by the year 2030. Restaurateurs, listen up.

At a wonderful brasserie in San Francisco called Café des Amis, a clangingly loud space with gorgeous tile floors, marble tables, and polished wood everywhere, the owners have cleverly set aside a room in the back, blanketed in "plush bordeaux-colored mohair" surfaces—exactly what the ALDA members were requesting: not a sterile banquet room in the back, but a fashionable space where they, too, can eat with dignity. It's a smart business move. There are plenty of youthful sixty-, seventy-, and eightysomethings who have money for dining out even if they can't hear as well as they used to.

The Sound of the Future

In 2010, Frito-Lay's SunChips brand of snacks launched what was claimed to be the world's first 100 percent compostable snack package. Immediately Frito-Lay started to receive complaints from consumers about the sound of the bags. Here was a snack food company trying to do the right thing for the

Earth, and consumers were complaining. In fact, they were more than mad. They were frustrated. No longer could the cheating dieter sneak a handful of chips in the middle of the night without rousing his spouse. Consumers created a Facebook page called SORRY BUT I CAN'T HEAR YOU OVER THIS SUN CHIPS BAG. The company responded that a loud compostable bag is "the sound of change." Then they pulled them off the market.

Actually, Frito-Lay was on to something. Amanda Wong and Charles Spence of the Crossmodal Research Laboratory found that people tasting Pringles (again, to assure that each crisp was exactly the same as the next) while hearing a recording of snack bags rated the crisps crisper when they heard the bags rattling than when they heard the canister of Pringles popping. SunChips could have parlayed this learning into some kind of response to the complaints, or they could have used this knowledge in the advance marketing of the compostable bag to head off the complaints in the first place: OUR EXTRA LOUD COMPOSTABLE BAG WILL NOT ONLY SAVE THE EARTH, IT WILL GIVE YOU MORE SENSORY STIMULATION.

The Rattle and Hum of Nettle and Plum

Much of the sound that influences our food behavior is "heard" without our conscious attention, perhaps because we're so used to constant low-level noise in the background of our daily lives that we unconsciously tune much of it out. Yet even without your knowing it, the music that a food retailer or restaurateur plays can influence what you buy.

One study showed that playing French music in a supermarket makes people buy French wine more than wine from other countries (in this study, specifically, Germany). Playing German music had the same effect, making customers buy more German wine than French. Yet fewer than 14 percent of the shoppers admitted that the type of music that was playing might have influenced their wine choice.

The tempo of the music that's played in a store can also influence your purchase decisions. Slow music makes grocery store shoppers slow down; that means they spend more time in the store, and this translates to more revenue per customer—a pretty awesome result from simply changing the radio station. We are so sure that we're in charge all the time, but in fact we can be manipulated like puppets.

Perhaps the most sonically challenged food venue is the grocery store. What are the signature sounds of a grocery store? I only recently considered this and I'm in the business. The soundscape at the front of the store is the ringing and dinging of the cash registers. The center of the store hums along with the cycling of freezer cases. The produce section sounds like, well, nothing. Retailers might consider adding the sounds of nature to influence sales of their fresh produce. If a clucking chicken can make bacon-and-egg ice cream seem eggier, wouldn't the pleasant sounds of the outdoors make produce seem fresher?

In fact, a few retailers are dabbling in this area. Safeway, a U.S. grocery store chain based in California, spritzes some of its produce displays with water. Just before the water starts, you hear the sound of gathering thunderclouds. Boom! With a crack of thunder, the "storm" hits the lettuce section. Produce grows outdoors and that's the sound that Safeway has recreated inside. It's an incredibly powerful reminder of where the food comes from.

I love casual Mexican restaurants and, in fact, just about any restaurant that offers fajitas. It's not just the flavor, it's the sound. They arrive at the table sizzling hot, *alive* on the platter; by contrast, burritos and enchiladas sound dead on arrival. Restaurants don't leverage this type of service enough. Their diners don't get to hear the satisfying sound of the raw meat hitting the grill. They miss the sizzle and pop of eggs on the flattop, bubbling in butter. I want more of these experiences to happen at the table, as at Korean barbecue restaurants, which offer raw meat that you cook at the table. It's a fantastic auditory, gustatory, and olfactory feast. You watch bright pink beef hit the grill with a sizzle, change color, release volatile aromas, and develop more umami with every second it remains on the grate.

hedonic *adjective.* 1: of, relating to, or characterized by pleasure 2: of, relating to, or characterized by hedonism.

hedonics *noun.* The branch of psychology that deals with pleasurable and unpleasurable states of consciousness.

Auditory cues are more than chemical reactions—they are the hedonic beginning of a food experience. And this hedonic experience should start when you're shopping, as at a farmers' market. There, repetitive electronic sounds are absent and in their place is the pleasant buzzing of human beings talking to one another. This has to be somewhat responsible for their resurgence.

Sensory Snack

Some people with misophonia can become completely unglued by the sounds of other people eating. While it's sound that is the offender, the disorder is most likely caused by a problem in the central nervous system. Hearing another person eating, chewing, or swallowing can alarm—even outrage—someone with this affliction.

Isolating Sound

Julian Treasure introduced me to an anechoic chamber, a special room designed to eliminate the echoic effect of sound, which bounces around and gets reflected back to us in normal situations. I decided I had to visit one so I could eat in it.

Luckily, there is such a room at the University of California, Berkeley, just across the Bay Bridge from my home. I e-mailed Professor Emeritus Ervin Hafter, who ran the Auditory Perception Lab at University of California, Berkeley, and explored aspects of hearing such as the spatial perception of sound and how noise reduction affects speech cognition. It turns out Hafter is a fellow foodie who had just returned from Bordeaux, so he was keen to talk about sound and food.

No one had ever asked to eat in Professor Hafter's chamber, but he was game. We took my bag of food in and Hafter closed the foot-thick outer door, and then another inner one. We were closed off from all the sound in the world, it seemed. Hafter told me to scream as loud as I could. I yelled.

Once. Twice. Again. The. Sound. Stopped. So. Abruptly. It. Seemed. To. Disappear. When I crunched into a celery stick, the sound was pure, clean, crisp, and beautiful. The first bite of an apple—in sound isolation from everything else in the world—punctuated the air with absolute clarity and an unmistakable imprint. If I were given the choice between eating an apple or a piece of chocolate in the chamber, I would choose the apple. Eating it was like making music.

I chewed a few other foods: a sourdough hard pretzel, a carrot, and a potato chip, recognizing each one's signature tones. When I left the chamber, I had a new appreciation for the stark beauty of the sound of food, often lost in the normal thrum of daily life.

We don't need to eat in anechoic chambers in order to appreciate the sensory thrill of sounds like that first bite of apple. We just need to listen more carefully to what our food has to say and find a new appreciation for its audio output. If only we could learn to take sensory pleasure from the sound of food, similar to the way we revel in the aroma or appearance of a dish. We might eat more healthfully if we fully realized that an apple delivers the type of sonic performance that you could never get from a bowl of ice cream. Even if all we did was pay more attention to the sound of the food we eat, we might take one less bite and enjoy it twice as much.

Taste: Hear Your Favorite Foods

YOU WILL NEED
   1 person to be the Eater (you), plus as many Guessers as you'd
      like, but this works best with a small number (1 to 5) of
      Guessers
   A handful of tortilla chips
   A handful of pretzels
   An apple
   A celery stalk
   A carrot

Any other loud foods you can assemble
A plate and an opaque cloth to cover the sound stimuli
Paper and pens or pencils

DIRECTIONS

1. Prepare your samples before the Guessers arrive. Put them on a plate and cover with a cloth so the Guessers can't see them.
2. Tell the Guessers you're going to eat several foods while they keep their eyes closed. Their objective is to try to identify the foods by their signature sound.
3. Have the Guessers close their eyes and keep them closed until you say it's okay to open them. (I like to make the Guessers hold up their right hand and promise not to cheat. It's no fun if you cheat!)
4. Eat each food as loudly as you can. Announce the samples as "the first food," "the second food," etc. You can spit them out to speed the exercise along.
5. When you are done, have the Guessers open their eyes and write down the foods they thought they heard.

OBSERVE

1. How many did each person get right?
2. What did you notice about the sounds?
3. Do you think you could do this with softer foods, like a pear or a cookie? Try!

6

How the Pros Taste

Sensory evaluation makes use of the remarkable virtuosity and range of the human senses as a multipurpose instrument for measuring the sensory characteristics of foods.

Michael O'Mahony, 1986

Roger and I walked into Delfina, our favorite neighborhood restaurant, on a random Monday night. We had arrived without a reservation, and the staff tried their best to accommodate us. We ended up at a table placed at the end of a long row of banquettes, those sofa-like seats that line a restaurant wall. Due to an odd wall configuration, my seat was adjacent to another two-seat table. If that party of two had added a third diner to their table, she would have sat exactly where I did. This thrust me into the midst of their discussion whether I wanted to listen or not. But once I started paying attention, the conversation was as irresistible as Delfina's olive-oil mashed potatoes.

The diners were two men, about twenty-five years apart in age. My first challenge was to figure out their relationship. When the young man spoke to the older one, he had an underlying level of disgust in his tone: they were father and son. For the next hour I—unavoidably—listened to their excruciatingly painful reunion after many years of estrangement. *You were not there for me. You would be healthier if you'd just let it go. You missed important events in my life.* It was like listening to a radio drama while dining.

I couldn't tell you what I ate that night. Yet I can vividly recall the intense family dynamics at the adjacent table. The drama had simply overridden my meal.

To explore the connection between attention and food, and how to get the most out of every bite, I turned to Herb Stone, who in 1974 cofounded Tragon, one of the world's premier taste-profiling firms. High profile companies hire Tragon to analyze a food or beverage when they're about to implement a change in ingredients, process, or suppliers, or when they experience a dip in sales or a clobbering by a new competitor. Tragon uses a technique called *quantitative descriptive analysis* to develop a sensory map of a food, using human tasters as compass and surveying tools. This map helps its customers understand the similarities and differences between their products and their competitors' products. After more than thirty years in the business of training thousands of panelists to become "multipurpose instruments," Stone has remarkably simple advice for anyone who wants to learn how to taste food more intensely.

"Pay attention," he says with absolute certainty. This is the single best way to get more from every bite. "Most people eat food unconsciously. It's sort of like, 'Look out, tongue: I've got stuff to get into my gut.'"

I agree. The simple act of focusing on your meal is critical to appreciating food. If you want to get more sensory input from your food, show it some respect. If you want to revel in your meal, it's important not to seat yourself next to an unfolding multigenerational family drama. To get more sensory input from your food, tune out the surroundings, juicy as they may be, and pay attention.

Evaluating Food Appreciatively

If you want to taste more of what you're missing, take a cue from wine tasters and apply their techniques to food. The instructions for savoring wine almost always begin with an evaluation of its appearance. The taster spends significant time simply looking at the wine before even smelling it. Is it cloudy? Is it clear? Deep dark red? Golden yellow or with a greenish hue? How saturated is the color? Swirl it around in the glass and watch how it streams down the sides. What does the appearance of the wine tell you about how it will savor?

If we put this type of effort into analyzing each meal we ate, our meals would progress more slowly and we would absorb much more sensory input. In the same way that the nutrients nourish our body, sensory input nourishes our psyche and makes a meal truly satisfying.

Consider your weekday morning breakfast, which you are likely to eat without devoting much attention to it. You may eat something similar every day and even completely divorce yourself from the food by fixing your attention on something else, such as a newspaper, a computer screen, or (shame on you!) the highway unfolding under your car. If you recognize these sensory-obfuscating behaviors, try something different tomorrow morning: pay attention.

Make eating breakfast your priority. Multitasking reduces the quality of your performance of all tasks. Before digging in, spend thirty seconds just looking at your breakfast. Is the color of each piece of cereal brown, golden, or ivory? Do the pieces vary in size and shape? Is the milk you poured over them thick and opaque or thin and watery? Is your coffee brown or black? Do the color and opacity change when you add milk? Did your toast brown evenly? Are the berries on your yogurt bright red or deep blue?

Start to view your quotidian breakfast as a sensory event. Observe a full sixty-second moment of alimentary appreciation before lifting a single utensil or eating a single bite. Put the newspaper aside for a day and simply pay attention to your breakfast and see how it changes the way you start your day. Visually inspect it as if you were seeing it for the first time. Smell it deeply before putting a bite in your mouth. Wait until you get to the office before checking your e-mail. If you must eat during the commute, find a car pool or use public transportation. Friends don't let friends eat and drive.

Human Being Versus Human Instrument

Eating mindfully is a way to make your experience with food more satisfying. For food professionals, eating mindfully is a job requirement. Being a professional means doing whatever it takes. For real estate agents, it means working every weekend. For doctors, it requires being on call. For lawyers it might mean representing a jerk of a client. For me, it means tasting food I hate.

When I taste something in a professional capacity, I consider it from two completely different perspectives. The first approach is to think critically about what I taste. The second—which may not even occur if I'm at work—is to consider whether I like it or not. For me to be successful at my job, making this distinction is critical.

You can appreciate the taste of something, but not really like it. You may appreciate the complexity of single malt scotch, but not want to drink it. Or you may appreciate the flavor and heat of a spicy salsa, but not enjoy the physical sensation (pain) involved in eating it.

As a professional food developer, I have to be completely objective. I am a human instrument. Personally, I despise eggs and I'm not really a huge fan of mayonnaise. Put the two together to make egg salad and you're likely to send me running out of the room gagging. The smell alone is enough to turn my stomach. Yet I have worked on many different projects where my professional responsibility was to taste mayonnaise or eggs or egg salad. Plain. Without spitting (because I don't believe in spitting). Over and over and over. In another context, Barb the human being would gag and scream "Blech!" at this affront to my taste buds. But Barb the human instrument is able to verbalize very objectively which of the egg salads I'm tasting has more acidity, which is sweeter, and which is saltier.

Consumers aren't professional food developers, so we don't make them taste things they don't like. But when we test the foods we develop at Mattson, we always have our target customers taste them to get their opinion of the early prototypes. For example, if we're developing a sports drink, we'll have athletes taste it. If we're developing a low-calorie meal, we might have weight-conscious women taste it. You don't want to try to improve egg salad for someone who will never, ever, in a thousand years order egg salad. Why? Because as a consumer who hates eggs, I would suggest improving egg salad by removing the eggs and the mayonnaise. You want to perfect your egg salad recipe for someone who loves egg salad.

We ask our target consumer both types of questions, depending on what we're trying to learn. The first type of question refers to the sensory attributes of the food: taste, smell, touch, vision, and hearing. We take an average of the scores from people who say they buy the product. If you'd never buy it, we don't want your opinions screwing up our data. One egg salad hater can skew our test results in the wrong direction. As a result of someone like me slamming our egg salad formula, we may make changes to it that make it less appealing to lovers of the product. When we ask about sensory attributes, we're not asking people whether they like the food, we're just asking what they think about the level of salt, sourness, or color. Notice the subtle difference.

Let's go back to the (hated) egg salad. When I'm at work in the food development lab, I am the proxy for consumers of egg salad in the early phases of development, before we're ready to give consumers a sample. If I ask myself a certain set of sensory questions, about the levels of certain tastes, textures, or aromas, I'll be much better able to communicate to my chef colleague than screaming "Blech!" and spitting it out in disgust. I may taste it and think that it's unpleasantly salty, or that the eggs are not quite firm enough, or that the color is too dark. To get at the sensory attributes of a product, we use a "just about right" scale that usually has an odd number of choices so you can anchor yourself with the "just about right" middle point (here, a 3) and indicate your score by marking "just about right" or choosing another rating on either side of the middle.

How is the level of salt in this egg salad?

1	2	3	4	5
Much too low	Somewhat too low	Just about right	Somewhat too high	Much too high

We ask "just about right" questions only for attributes that we have the ability to adjust. For example, if we're working with a certain egg farmer who raises a certain breed of chickens, we might not want to ask about the color of the egg yolk if there's no way we can change it.

Finally, after we've captured the sensory attributes, we may ask consumers: How much do you *like* this product? The food industry standard is a 9-point scale ranging from "like extremely" to "dislike extremely." If I were to taste egg salad, I'd score it very low in terms of liking. But I can't very well taste a prototype in our food lab, turn to my chef or food technologist colleague, and say, "I dislike this extremely. Please change it."

The following is an example of a typical scale we use to let the target consumer note his personal preference. The 9-point scale is referred to as *hedonic*, a word that shares its Greek origin with hedonism, for obvious reasons: What's more hedonistic than a food that has a "just about right" level of fat, salt, and sugar? There's a clear midpoint that allows you to anchor yourself with a neutral middle score (5) and adjust your score up or down from there. Note that the "best" score on the "just about right" scale is in the middle while the "best" score on the hedonic scale is at the far right hand (or highest number).

How much do you like this egg salad?

1	2	3	4	5	6	7	8	9
Dislike extremely	Dislike very much	Dislike moderately	Dislike slightly	Neither like nor dislike	Like slightly	Like moderately	Like very much	Like extremely

Developing a Lexicon of Attributes

To critically analyze a food, start by asking yourself if all five of the Basic Tastes *should* be present. For some foods, the answer is yes. For others, the answer is no. And sometimes it's not so cut-and-dried.

Let's move from egg salad (please) to something else. Like chocolate. Imagine yourself trying to make the decision between two different milk chocolate bars. You sample one, then the other. How do you decide which one to eat? Here's how I would proceed.

The first sense you use in the evaluation of food is your sight. Milk chocolate bars might have a range of brown colors and a range of shine or gloss, and if the chocolate was not tempered or stored properly, there might be a slight bloom on it. Chocolate bloom is usually harmless, even though the resulting white coating looks a bit like mold and is considered a negative appearance quality.

Next we move to the second sense, taste. Should all five Basic Tastes be present? Chocolate should be sweet. But how sweet? Should it be sour? Bitter? What about salt? And umami?

Some premium varietal chocolate (made from a single type of cocoa bean) has a slightly fruity note, which may come across as sour. Some milk chocolates have a sour-milk note that gives them a certain distinctive character. But for something sour, you'd probably choose a Jolly Rancher or Life Saver. If you wanted something chocolaty *and* sour, you might reach for Raisinets, in which sour is entirely appropriate. Regular chocolate has a very low level of sourness. But since sour is evident in chocolate, it goes on the attribute list.

Bitter may not be a taste you want in many foods, but in chocolate a touch of bitterness can lend depth and complexity, which chocolate that lacks bitterness doesn't have. Bitterness is inherent in cocoa beans, so it's likely to be there anyway. We're just talking about the sensory aspects, not which type of chocolate (milk or dark, bitter or not) is better. We'll get to preference later. It's probably good to have a little bit of bitterness in a chocolate bar. How much? Just about the right amount. Bitter, then, goes on the list of attributes.

Salt is another flavor you wouldn't usually associate with chocolate, but salt gives chocolate a counterpoint. Think of the perfectly balanced five-pointed Taste Star. If our chocolate is sweet (the first point) and slightly bitter (the second point), adding salt will give it another point (the third point). Each counterpoint adds contrast. And if they're at a level that's *just about right*, they add balance.

Back to salt. Since our chocolate tasting is for plain milk chocolate bars, not sea salt chocolates or salted caramels, let's agree that salt may be present but it probably shouldn't be obvious—that would mean it was out of balance. Salt goes on the list.

Umami is another taste that usually isn't present in chocolate. I have tasted chocolate with umami, which changes the taste dramatically, but it has to be at a very low level or it sticks out like a sore thumb. Trendy chocolate bars with bacon in them have the Basic Taste of umami as well as salt and the aroma of smoke, for an interesting twist. But for this exercise, we'll leave umami off the list of attributes for milk chocolate.

Now we move from taste to texture, a key component of chocolate bars. This requires a completely different sense. Now you're going to have to put taste aside and rely on your sense of touch. In the food industry, we refer to the texture of foods in your mouth as *mouthfeel*—how a food feels in your mouth. Should the mouthfeel of a perfect chocolate bar be gritty, grainy, smooth, crispy, crunchy, creamy, thin, fatty, or chewy? Should the chocolate linger on your tongue or disappear quickly? Since you're evaluating a regular milk chocolate bar, without nuts or other ingredients, let's say that it should be smooth and creamy with a fatty texture that lingers on your tongue but doesn't feel waxy. All of these attributes go on the list.

Next sense: smell. Now you're ready to consider which aromas you expect in a chocolate bar, the same way you consider which of the five Basic Tastes should be present. Try inhaling its aroma in a long, sustained slow breath, which is more effective at detecting odors than short repetitive sniffs. The chocolate may have a subtle aroma when you nose-smell it, but probably not much, because the volatile aromas in chocolate aren't released until the chocolate is heated or melted—both of which happen in your mouth. Once you've gleaned what you can from nose-smelling, you'll put the chocolate bar in your mouth to mouth-smell it, which is when the aromas come alive.

Coming up with the terms you use to describe the two chocolate bars is called *developing a lexicon*. At Tragon, the taste-profiling firm, its trained consumer panelists choose the terms in the lexicon. To create these terms, they taste a broad, heterogeneous range of the food being worked on: cookies, wines, chocolates. For this chocolate tasting exercise, it would mean assembling lots of milk chocolate bars—from cheap to expensive—to make sure you've covered all the aroma ground of chocolate.

In order to train its panelists to detect a certain attribute, Tragon might give the panelists a reference food—an example of something that has the prototypical flavor they're trying to identify. Tragon and other sensory testing firms are trying to use humans as tasting instruments; in order to measure something with an instrument, you first have to calibrate it. By tasting or smelling a reference, panelists are able to calibrate their appropriate sense to what that characteristic refers to. In order for a panel of tasters to rate the level of rancidity in a food, for example, they all need to agree on what rancidity is. The best way to do this is to smell something rancid, agree it's rancid, and pass it around the room until everyone is calibrated on what rancidity smells like. Then, move on to the next reference.

In our chocolate tasting, the reference for vanilla might be a sniff of a vanilla bean or a sniff from a bottle of vanilla extract. The reference for the attribute "nutty" might be a selection of roasted mixed nuts. Of course, you'll want references for the Basic Tastes, too. Table salt is usually the reference for salt, and sugar the reference for sweet. Citric acid makes a great reference for sour, but if you don't work in a food lab that stocks it, you can use ascorbic acid, the technical term for vitamin C. You can buy vitamin C capsules (not tablets) from a health store and open them to taste the powdered contents—a pure reference for sour. Usually quinine, the bittering agent in tonic water, is used as a bitter reference, but again, it's not easy to find. I like using powdered caffeine, which you can make by pulverizing over-the-counter pills such as NoDoz, and dissolving it in hot water. But be careful. Even a tiny taste of caffeinated water is very bitter and more than a tiny bit will keep you up at night. A standard reference for umami is monosodium glutamate, which you can buy at the grocery store, sold as the seasoning brand Accént.

In a typical milk chocolate, the aromas would include milk, cocoa, and roasted smells, since chocolate (cocoa) beans are roasted. But there may be some other characteristic flavors such as cooked and caramelized milk, sour milk, coffee, smoke, leather, fruit, nuttiness, or earth. The reference samples for these would include caramel, coffee, liquid smoke, real leather—you get the point.

Lexicon and References for Milk Chocolate

Descriptive Term	Reference Food for Training Tasters to Identify the Flavor
Milky	Evaporated milk
Bitter	77% cacao cocoa powder
Stale (rancid)	Oxidized old butter
Rum	Dark rum
Vanilla	Bourbon vanilla beans
Nutty	Roasted mixed nuts, lightly salted
Astringent	Drying sensation from tasting cocoa powder
Sweet	Granulated sugar

Excerpted from J. Kennedy and H. Heymann, *Projected Mapping and Descriptive Analysis of Milk and Dark Chocolates.*

Last, you'll want to consider the sound of chocolate. If the bar snaps loud and hard, it likely has been tempered, or melted, properly. When chocolate is tempered poorly, the snap sound is much less pronounced.

Now that you've considered the basic tastes, aromas, visuals, sound, and texture, you can start to make judgments about the two chocolates. It's likely that one of them is creamier, one is more bitter, one is milkier, and one is smoother. In making these critical judgments, you are conducting a sensory evaluation.

After you've evaluated each individual attribute, you can move on to asking which chocolate you prefer. This exercise should be easy for you, but the interesting thing to ask is *why* you chose the one you did. Do you gravitate toward more bitter or sweet foods? Do you like very creamy (fatty) foods because of the mouthfeel on your tongue? To really figure these things out, complete the Sensory Evaluation of Milk Chocolate Bars exercise at the end of this chapter. Yes, it will be rough, but someone's gotta do it.

Man Versus the Machine

The type of sensory analysis that's done at Tragon and other sensory testing companies is very different from the type of taste testing we do at Mattson. Tragon trains human panelists to approach food in an unnatural way, by design. They want consumers to calibrate their sensory experiences with reference foods and other tasters, who in essence function as analytical machines. The companies don't use actual machines. Humans will taste something for free in many cases (chocolate, for example) and for very little money in other cases (egg salad, for example, although I'd charge more than most). The thirty- to seventy-five-dollar cost of hiring a consumer to taste samples is much less expensive than putting the same number of samples through a machine such as a gas chromatograph. The results are also more useful, as humans can detect which compounds other humans will also like or dislike. Machines, on the other hand, may be able to detect more compounds in a sample, but those compounds may not necessarily be the ones that people care about. Lastly, humans are mobile, and you can find them just about anywhere, which isn't true for expensive analytical machines.

At Mattson we test food on humans, too, in the course of improving prototypes during development. First we ask about the sensory characteristics, then we ask how much tasters like the food. Most importantly we ask them if they would buy it. Because you can have just-about-right levels of everything in a dung-fire-grilled squirrel steak, but if a consumer isn't going to buy it, the testing is moot.

Strongest Imaginable Sensations

It's worth noting here that trying to calibrate perceptions of taste is frustrating, because any two (or two hundred and two) people who taste a single food will experience it in different ways. Every human being lives in his own sensory world and about one-quarter of the population is made up of HyperTasters, who may experience a food as being three times as intense as Tolerant Tasters do. The human range of olfactory perception varies tremendously, especially as we age.

Companies like Mattson and Tragon test food with people who actually purchase the product, usually with rabid fans of the type of food they're developing or testing. Tragon puts consumers through a battery of tests to prove their sensory skills. In other words, if you are a Coca-Cola drinker and Tragon is testing soft drinks, it will give you taste tests to make sure you can tell the difference between Cola Sample A and Cola Sample B. After a round of testing, says founder Herb Stone, about 30 percent of the population drops out of the running.

"They actually defy statistics," he says, meaning that these (most likely) Tolerant Tasters score worse than chance on the taste tests. "By chance alone, they should get fifty-fifty. They don't. You get some people with twenty percent, thirty percent, and at the other end, you get people at ninety percent correct." Even though these people are regular consumers of the product, 30 percent of them (and of consumers of any product) can't identify it in a blind tasting. This is hugely significant for makers of private-brand foods and drinks, because they have to market to these Tolerant Tasters without being too transparent. YOU CAN'T TELL THE DIFFERENCE ANYWAY, SO BUY OUR PRODUCT! isn't going to win any advertising awards or motivate consumers to buy. But for 25 to 30 percent of the population, it's true.

One of our clients hired us to conduct some testing on its salad dressings versus their competition. We recommended doing blind taste testing, in which testers evaluate the samples without knowing what brand they are tasting. Our client's brand of salad dressing didn't score as well as its private label competitors' did. Yet after the tasting, we asked people to name the brand of this dressing they'd be most likely to buy. Just like well-trained consumers, they rattled off the leader's brand. Behold the power of marketing.

Academic researchers in the field of sensory science, such as Linda Bartoshuk of the University of Florida and Barry Green from Yale University, find that the descriptive analysis method isn't accurate enough. They argue that the differences between people are still too great to lump together and average out. Instead, they've come up with a new way of measuring sensory experience. It's called the Generalized Labeled Magnitude Scale, or gLMS, and it's now the standard for sensory testing in academic research. The gLMS scale is special because of its upper limit, which isn't a 9, a 100, or a description such as "extremely intense," all arbitrary designations that could mean one thing to you and another to me. Instead, the top of the scale—the highest score—is set by the taster.

Roger and I used the gLMS method of scoring foods when we visited the Center for Taste and Smell in Florida. We were asked to set our upper limit—our highest score—by coming up with the "strongest imaginable sensation of any kind" that we had each experienced. They gave us a few examples. Many women rate childbirth as their strongest imaginable sensation. Those unlucky souls who have passed kidney stones often choose this as their strongest imaginable sensation. Roger remembers the very moment during a dental procedure when his dentist hit a nerve. He asked the doctor for more anesthetic, but was told that he had already maxed out, so there was no way to up the dose safely. That five-minute hell was Roger's strongest sensation. I had a harder time with mine and realized as a result just how cushy my life has been.

The next thing the gLMS administrators did was ask us to rate the brightness of the sun, and other things that we all experience. This gave them a couple of data points so they could "normalize," or calibrate, our scores from this point forward.

Once we'd established the highest point on our scoring line, we were asked to start evaluating food. It is an odd thing to consider for the first time: how strong is the sensation of sweetness in a piece of chocolate compared with five minutes of anesthetic-free dental work? It takes some time to figure the scale out. We eventually got the hang of it as we rated everything from the five Basic Tastes to popcorn, lasagna, and grape jelly.

Hot, Hotter, Hottest

If you are going to evaluate more than two foods at a time, it's very important to start off with the right one. Let's use two foods as examples. First: salsa.

If you were conducting a salsa tasting, you'd want to sample salsas from mildest to hottest, because the heat from capsaicin (the compound that gives chile peppers their kick) tends to build up over multiple mouthfuls. With each mouthful, you also become more conditioned to the flavor of the salsa, so you want to be sure you're making your judgment after the first bite or two. Malcolm Gladwell's book *Blink* describes exactly what I do when I taste something in the lab. I try to make an initial *Blink*-style judgment about it: I go with my initial gut-based instinct in the first instant my senses are activated. I file that away as I continue tasting, and then see if anything changes as I keep eating. I need to take only one or two bites to make a comprehensive judgment about what I taste—thankfully, or I'd weigh 300 pounds—but trust me, it has taken decades to develop this skill. Over the years I've come to trust my *Blink* assessment as the most accurate.

For our second example, let's imagine that you are tasting four or five different cups of black coffee. Coffee has a famously bitter taste that also builds up over time, so it's best to start with the least bitter of the samples, then move on to progressively more bitter ones. Save for last the thick, heavy brew that will put hair on your chest. The same holds true for just about any category of food or beverage. The basic rule is to move from mildest to increasingly intense.

While the chefs and food technologists at Mattson know and practice this concept, they can't prevent themselves from biasing me. When I am called into the food lab to taste a prototype, the first thing out of the developer's mouth is usually, "Sample A has more . . ."—at which point I have to put my hand up and tell him to stop. Even though I'm acting as a human instrument, I'm unfortunately incapable of escaping human nature, which is to be biased by what you are told. So if Lin or Gorski tells me that Sample A is more salty than Sample B, it's entirely likely that I will perceive this. I try very hard not to let this type of information influence me, but that's really difficult, so I prefer not to hear it at all. I feel the same about reading wine tasting notes before trying a wine, reading a movie review before going to the theater to see it, and hearing a critic's views on a new restaurant before I can form my own, unbiased opinion.

The Dilution Solution

In the case of some really intense foods, such as hot sauce, we take another step when tasting. To ensure we can evaluate the full spectrum of tastes and flavors, we dilute the food with water. Consider the hot sauce sriracha (known to its diehard fans as "rooster sauce" because of the iconic rooster on the label of the best-known brand). Its label is designed to look Asian, though it's made in America, and it's ubiquitous at Asian restaurants. This intense, bright red sauce is thick, vinegary, and dominated by a deliciously fresh chile top note. It's also scorchingly hot. A tiny dot on your tongue will put your taste buds out of commission for a few minutes. To avoid this, we dilute a small amount of it in water and then taste the sriracha-flavored water. By doing this we can swish the sample around to experience all of the volatile aromatics. And because it's less hot, we can take a second or third taste. Professionals who taste alcoholic spirits use this same method. Because the alcohol level of most spirits is so high, it's hard to distinguish subtle flavors, and it can be unpleasant to swish them around in your mouth—not to mention inebriating. By diluting spirits with water you can detect more than just the burn of the alcohol. Some scotch aficionados will be likely to ask that you dilute their scotch with a bit of water, because they know how to get more flavor enjoyment from their drink of choice. This technique works even for foods you wouldn't imagine diluting. Like chicken.

We at Mattson were working for a fast food restaurant chain whose research and development department had spent about six months developing their own twist on a very famous fried chicken sandwich. This famous fried chicken sandwich is, frankly, a work of art. It's a boneless, skinless chicken breast, battered, then breaded with a simple blend of flour and spices. It's fried to golden tenderness—not crispness—that results in a firm but juicy bite. Served on sweet, squishy white bread buns with dill pickles, the sandwich is a perfect balance of sweet, sour, salt, and umami. To me, it's a crave-able, truly American indulgence right up there with apple pie, hot dogs, and bourbon-cornflake ice cream.

Our client's goal was to develop something similar that they could offer in their restaurants for less money than the competition. They had been relatively successful in creating the sandwich, given the amount of time they'd put into the effort. Yet they were still about five percentage points away from parity preference, which is when consumers like both products the same. We use this as a measure of success when we're trying to match, or reverse engineer, a food.

We frequently get requests from clients to "knock off Heinz ketchup" or "create an Oreo cookie clone." Retail grocery stores, for example, might want to create a knockoff of a classic product so that they can sell it under their own store brand, usually at a lower cost. These projects are virtually impossible. As technologically advanced beings, we want to believe that there is a magical black box into which you can feed anything—a wine, a pharmaceutical, a voice, a tomato-based condiment—and it will spit out the complete instructions for how to re-create the product. Because you can do this type of matching with paint chips, people assume it must be possible with food.

In fact, reverse-engineering is always harder than you anticipate, because there are so many factors affecting the way a food tastes. Even if you know the fat, carbohydrate, and protein content of a food, the ingredient list, and its physical specifications (viscosity, pH, Brix, salt, and so on), that's still not enough. The simple difference between tomato varieties can affect ketchup flavor significantly. At least for now, there are too many missing inputs to effectively create a perfect clone. The best method we have is trial and error. We whip up a batch of food, taste it as a team, then revise the samples and repeat the whole process—again and again.

At Mattson we won't take on a project with a perfect match as the goal; it's simply too difficult to accomplish in a reasonable amount of time for a reasonable amount of money. We *are* willing to take on a project with the goal of matching the degree of liking or achievement of parity preference. Consumers may recognize that Ketchup Sample 308 that we've created is not exactly the same as Ketchup Sample 215, but they may like them both equally well. We are usually confident we can create an equally liked food, and we frequently do.

Back to the chicken. Our client was very, very close to achieving parity preference on the fried chicken sandwich they had developed to go up against the competition, but they weren't quite there yet, even after rounds and rounds of trial-and-error development. Before they'd come to us, they'd also sent the product they were trying to copy for gas chromatography/mass spectrometry analysis. They'd analyzed the physical specifications of the product and hired a trained panel to conduct a descriptive analysis. But they could not close the gap between their sandwich and the competition's original. So they asked us to help.

After much discussion about untried methods, I decided to approach this challenge the same way we evaluated the hot sauce. Others who had tasted the product at full strength had missed key flavor notes. As a consumer of the gold-standard product, I knew that the characteristic flavors are very subtle. Perhaps we could use the same dilution methodology to decipher what we were missing. Perhaps by diluting the ingredients in the sandwich, we could spread them out—like smearing paint across a canvas—to see what was in the trail. It was worth a try.

The first thing we did was taste the chicken sandwiches side by side. Then we tasted the fried chicken fillet without condiments or bun. Then we ran three additional taste tests using our chosen methodology.

First we put the battered and breaded fried chicken breast patty in a commercial blender with 200°F deionized water and whirred it to smithereens. Chicken smoothies, anyone? Then we filtered the smooth liquid puree through cheesecloth so that we were left with the very essence of the fried chicken breast fillet. What we had done was somewhat of a distillation, a tea-style steeping, or an extraction . . . of fried chicken.

Then we did the same for the chicken scraped clean of its breading. Scraping chicken breasts of their breading is the type of thankless task that interns at Mattson are asked to do. And just to make sure we'd covered all our bases, we whirred up a chicken breading smoothie extraction: just the scraped-off breading, none of the chicken. I know you're salivating now.

This exercise turned out to be amazingly productive. After tasting fried chicken, naked chicken, and chicken breading in their natural states, we were almost sure we had identified the differences between the two sandwiches.

But once we tasted the extracted broths, we were absolutely certain. Our client's chicken breast was too high in umami and certain herbs, had too much of an oily mouthfeel, and retained too much oil flavor from the frying process. These characteristics were obvious to us because of the method we'd used. And, no, my love for that chicken sandwich has not diminished even after I deconstructed it in this horrific way.

Sensory Snack

I asked Clos du Bois winemaker Erik Olsen why I couldn't find a low-alcohol wine that didn't have anything else added to it (such as sugar, flavors, or sweeteners). He told me that low-alcohol (and nonalcoholic) wines just don't taste right. Without the natural sweetness and tactile burn that alcohol brings, wines taste unbalanced when you go below 8 or 9 percent alcohol. The sugar, flavor, or sweetener is there to get it back in balance.

Tasting with Your Mouth Shut

I once had a brilliant client who was a bit volatile. Actually, that may understate his level of excitability. He was known to jump up from his chair and shout things out during meetings. He yelled and screamed when he was unhappy. And when he was pleased, he was effusive. These personality characteristics are not unusual in an entrepreneur, but they're terrible during a food tasting meeting, or a *cutting*, as we call it.

The way a cutting is supposed to work is that everyone at the meeting tastes the sample, everyone writes down their comments before anyone says anything, and then—once everyone has had a chance to make up their minds about the food—we share. The problem with this client was that the minute he tasted something he wasn't crazy about, he would blurt out, "UGH! That is DISGUSTING." Or, "That's a fat BOMB. Who in their right mind would eat that?" or "That is the WORST thing I've ever tasted."

Now, imagine you're in a cutting with him and you really like the thing you are tasting. What if it was your idea? Worse yet, what if you had created the prototype? This client's negative comments could sway your opinion or at the very least your desire to verbalize your opinion. During one meeting, this exact scenario happened and I lost my temper with him. I couldn't help myself.

"Can you please keep your comments to yourself until everyone has a chance to taste?" I asked, annoyed.

"Yes, but that appetizer is WRONG!" he spewed.

"No, seriously. You have to keep your mouth shut. When you shout out something while we are still tasting, it's like you just let out a big fart. It's impossible to be objective about the food we're eating while a bad smell is wafting through the room, hanging there, influencing the way we taste the food from that point forward. Your negativity influences our decisions."

"Really?"

"Yes, really. Keeping quiet during the initial tasting is the same protocol we follow when we're testing prototypes with consumers. It shouldn't be any different with us, the professionals."

"You're absolutely right," he said. He totally took this criticism to heart. In fact, he's since become quite adept at sensory analysis. And I'm happy to say he never again "farted" on my prototypes.

Cleansing the Palate

There are times when you want the tastes and flavors from the foods you're eating to blend together. I like to dip the corner of my grilled cheese sandwich into tomato soup, for example, because I love the taste of cheese and tomato together in my mouth. The salt and umami tastes in both foods come together synergistically for an explosion of savoriness. When you're trying to critically analyze foods, you want to do just the opposite. You want to clear your mouth of the taste you just experienced and ready it for the next one.

Professionals cleanse their palates to minimize adaptation. Remember the competitions in which tasters are asked to make judgments on dozens and dozens of wines? According to the concept of adaptation, they'll be less sharply attuned to each additional taste. This holds true for both taste and smell. Consider walking into a smelly room, perhaps, in which someone has just steamed a bunch of sulfury cauliflower. Or it could be a lovely-smelling room where bread is baking in the oven. Take your choice. In either situation, your brain would short itself out if it kept pinging you with the smell information it is perceiving. It might go something like this inside your brain: toasty, sweet, buttery, toasty, sweet, buttery, yum! Yum! Toasty, sweet, buttery, yum! And so on.

In order to save itself from exploding, your brain adapts to the input instead, and eventually you don't even notice that the room smells. That is, until you leave, go outside, and return again to the room, refreshed. That's what cleansing your palate is also meant to do: refresh you so that you won't shut down from too much sensory input.

The first time you walk into that smelly room, you experience what I call your *virgin* smell: your first experience with it. The difference between sensory stimuli and losing your own innocence is that you can return to the virgin smell state. You can leave a stinky room and refresh your nose, and you can cleanse your mouth to refresh an adapted palate.

Many things work for cleansing your palate, but food-tasting professionals generally use saltines washed down with filtered water. Saltines contain salt, which is a functional ingredient in the dough, but you can find saltines with unsalted tops, which are what's commonly used in the industry. Saltines are the best choice for cleansing the palate because they don't have strong, characteristic flavors like butter, yeast, or nuttiness, which are evident in most baked goods. You'll often find saltines at wineries that cater to the type of discriminating taster who practices palate cleansing.

I sometimes like to use sparkling water to wash down my saltines, because it feels like the bubbles cut through the fat of some foods better than still water. Either way, a bland cracker and a big swish (or two) of water will help prepare your tongue for the next food. It would be extraordinarily difficult to distinguish between Hot Salsa Sample 3 and Hot Salsa Samples 4 and 5 without cleansing your palate. Cleansing between samples is important in order for you to tell where the tactile heat of one sample starts, when the next one kicks in, and when the burn from each is extinguished on its own. Sometimes it even makes sense to wait a few minutes between sampling.

When I was being tested at the University of Florida's Center for Smell and Taste, I was given a cup of coffee grounds to sniff between tastings to cleanse my sense of smell. I found this distracting, because coffee is neither mild nor neutral. Some other food professionals I've talked to suggest sniffing your own skin, such as the back of your hand. Both of these methods refresh your sense of smell with something else, so they are akin to leaving a room that smells of bread baking. It's easier in most cases to simply remove the food from your vicinity and take a deep breath of fresh air.

Before doing any of the exercises in *Taste*, you should start by eating a saltine and drinking some water. Then you're ready to taste like a pro.

Taste: The Concept of Adaptation

YOU WILL NEED
>1-tablespoon measure
>Sugar (any type)
>2 glasses
>1-cup measure
>Warm water
>A spoon
>Saltine crackers for cleansing your palate
>Cold water

DIRECTIONS
>1. Measure 1 tablespoon sugar and pour it into the first glass.
>   Add 1 cup warm water and stir briskly to dissolve the sugar.
>2. Measure 4 tablespoons sugar and pour it into the second glass.
>   Add 1 cup warm water and stir briskly to dissolve.

TASTE
>3. Taste the mildly sugary water first. Note how sweet it tastes.
>   This is your virgin taste.

4. Now taste the extra sugary water. Note how sweet it tastes.

5. Without cleansing your palate, go back and taste the first glass. You'll notice now that the water tastes less sweet than it did when your palate was fresh. That's because your palate is no longer virgin. You can go back and forth a few times to see how hard it becomes to distinguish sweetness levels.

6. Now cleanse your palate with a cracker, rinse your mouth out with cold water, and wait a few minutes. Then go back and taste the first cup of sweetened water. You'll see that cleansing your palate really does help alleviate adaptation.

Taste: The Dilution Solution

YOU WILL NEED

One type of hot sauce such as sriracha, Tabasco, or Frank's

4 cups

Tablespoon measure

Water

Whiskey, bourbon, or some other complex brown liquor

Tasting spoons

Saltine crackers for cleansing your palate

DIRECTIONS

1. Pour a small amount of hot sauce into each of two cups. To one, add a few tablespoons water and stir.

2. Pour a small amount of liquor into the two other cups. To one, add a few tablespoons water and stir.

3. Taste the straight hot sauce first. Try to describe what you're tasting.

4. Cleanse your palate with crackers and water, then wait a few minutes.

5. Try the hot sauce diluted with water. See if you can detect more of the flavor of the hot sauce when it's diluted.

6. Cleanse your palate with crackers and water, then wait a few minutes.

7. Repeat the exercise with the liquor.

Taste: Sensory Evaluation of Milk Chocolate Bars

YOU WILL NEED

2 different brands of milk chocolate bars: give each one a letter code. One should be "A" and the other "B."

Saltine crackers and water for cleansing your palate between tastings

1 printed copy of the Sensory Evaluation and Preference page for sample A for each person who is tasting

1 printed copy of the Sensory Evaluation and Preference page for sample B for each person who is tasting

DIRECTIONS

1. Eat a saltine and drink some water to cleanse your palate.

2. Fill out the Sensory Evaluation form for A first.

3. Fill out the Preference form for A second.

4. Repeat all steps for B.

PART 1: SENSORY EVALUATION

TASTER NAME: _____

SAMPLE: (CIRCLE)   A   B

How is the **brown color** of this chocolate bar?

1	2	3	4	5
Much too light	Somewhat too light	Just about right	Somewhat too dark	Much too dark

How is the **shininess/glossiness** of this chocolate bar?

1	2	3	4	5
Much too flat/matte	Somewhat too flat/matte	Just about right	Somewhat too shiny/glossy	Much too shiny/glossy

Do you notice any **bloom** on this chocolate bar?

1	2
Yes	No

How is the "snap" **sound** of this chocolate bar when you break it?

1	2	3	4	5
Much too loud	Somewhat too loud	Just about right	Somewhat too low	Much too low

How is the level of **sweetness** in this chocolate bar?

1	2	3	4	5
Much too low	Somewhat too low	Just about right	Somewhat too high	Much too high

How is the level of **tartness** in this chocolate bar?

1	2	3	4	5
Much too low	Somewhat too low	Just about right	Somewhat too high	Much too high

How is the level of **saltiness** in this chocolate bar?

1	2	3	4	5
Much too low	Somewhat too low	Just about right	Somewhat too high	Much too high

How is the level of **bitterness** in this chocolate bar?

1	2	3	4	5
Much too low	Somewhat too low	Just about right	Somewhat too high	Much too high

How is the level of **chocolate aroma** in this chocolate bar?

1	2	3	4	5
Much too low	Somewhat too low	Just about right	Somewhat too high	Much too high

How is the level of **roasty aroma** in this chocolate bar?

1	2	3	4	5
Much too low	Somewhat too low	Just about right	Somewhat too high	Much too high

How is the level of **dairy/milky aroma** in this chocolate bar?

1	2	3	4	5
Much too low	Somewhat too low	Just about right	Somewhat too high	Much too high

How is the level of **fruity aroma** in this chocolate bar?

1	2	3	4	5
Much too low	Somewhat too low	Just about right	Somewhat too high	Much too high

How **smooth** is the texture of this chocolate bar?

1	2	3	4	5
Much too smooth	Somewhat too smooth	Just about right	Somewhat too grainy	Much too grainy

How **waxy** is the texture of this chocolate bar?

1	2	3	4	5
Much too fatty	Somewhat too fatty	Just about right	Somewhat too waxy	Much too waxy

How **creamy** is a mouthful of this chocolate bar?

1	2	3	4	5
Much too low	Somewhat too low	Just about right	Somewhat too high	Much too high

PART 2: PREFERENCE

How much do you **like** this chocolate bar?

1	2	3	4	5	6	7	8	9
Dislike ex-tremely	Dislike very much	Dislike moder-ately	Dislike slightly	Neither like nor dislike	Like slightly	Like moder-ately	Like very much	Like ex-tremely

Which sample do you prefer? SAMPLE: (circle)   A   B

Why?

OBSERVE

1. Compare your ratings with those of your fellow tasters. How do they differ?
2. Why did you like your preferred sample?
3. Do you think this preference extends to other foods and beverages?

7

From Womb to Tomb

My friend Sally has two beautiful, healthy, vivacious blond girls she adores, but anytime her husband starts a discussion about having a third child, Sally immediately quashes his hopes. While she wouldn't mind a third child, she has her reasons for stopping at two.

"I didn't enjoy being pregnant," she says of the unrelenting morning sickness she experienced during both pregnancies. "It was so bad. With Kathryn it lasted five months. With Barrett, my second, it started at seven weeks, lasted the whole pregnancy. It never went away. I threw up, actually, the day I gave birth." Even the thought of being pregnant again makes Sally cringe with a visceral physical reaction. "I could never be pregnant again," she says definitively.

Not surprisingly, Sally had trouble eating while pregnant. The mere sight of some foods, such as raw chicken, would make her vomit. Tomatoes, which she usually loves, would send her running to the bathroom. Her obstetrician put her on the antinausea drug Zofran, which is used to help cancer patients cope with the side effects of chemotherapy.

"It was worse in the early stages," Sally remembers. "I would throw up throughout the day. Probably by the time I got into my second and third trimester, I would maybe throw up once or twice a day, so it wasn't as bad." Not as bad as what? Chemotherapy?

Sally and I have been friends since college. We both live in the San Francisco Bay area, and we get together every now and then to catch up. So when I learned an interesting fact about how taste preferences develop, I went to see her.

I started by asking Barrett, Sally's youngest, what her favorite food is.

"Pasta!" she shouted out without hesitation. "I like salt on my pasta! A lot." Later Barrett said, "I also like chips and cheese." At this point, her big sister Kathryn told me about Barrett's diet. "Normally, all she has for lunch is chips and cheese." Sally confirmed this. Sally said she would often make the girls sandwiches, which she'd serve with potato chips, string cheese, and fruit. Barrett would completely ignore the sandwich and fruit, eating only the salty chips and cheese.

"If I would let them, they both would load up the salt on their food. It's to the point that it has to be supervised," says Sally about her daughters' intake of sodium.

Next I talked to Darren, who is Sally's husband and Barrett and Kathryn's dad. I asked him to describe the use of salt by the women in his family.

"It's out of control," he said promptly. "They put salt on everything. Barrett just asked for salt for her corn. She puts salt on her pasta," he said, reconfirming what Barrett had told me earlier. "From very early on, Kathryn would always put salt on her food. 'Pass the salt,'" Darren said, mimicking his daughters' refrain at the dinner table.

Darren just assumed that his young daughters were mimicking their mother's behavior; they salted their food because Sally salted her food, and this was now an ingrained behavior. I was fairly sure there was a different explanation.

Researchers in France have shown that maternal vomiting affected pregnant rats' offspring. When I first read this, I imagined tiny glowing pink rats, with swollen pregnant bellies, draped over teacup-size water bowls, waiting for their morning sickness to pass. But in fact the researchers had simulated the effects of vomiting by giving the rats polyethylene glycol, which made them dehydrated and, more important, made them crave salt. Sure enough, when the little pink baby rats were old enough to feed themselves, the ones from mothers who had craved salt during pregnancy were themselves saltaholics.

Researchers also tested four-month-old human babies for their preference for salty water. Sure enough, just like the rats, the babies whose mothers had suffered from vomiting when pregnant liked salt more than babies whose mothers didn't get sick. If women are sick and dehydrated when pregnant (and ostensibly craving sodium, which they lose when they vomit), their fetuses experience this craving, too, and the babies are then born with preprogrammed preferences for salt. This preference for salt continues well into adulthood. When researchers tested older children, their saltaholic tendencies had stuck around, well after they were weaned from their mothers and presumably fully hydrated.

If love for salt can develop in utero, what other food preferences are we born with, and which are learned?

Taste the Truth

Our likes and dislikes of the Basic Tastes are connected, beautifully, with our built-in will to survive. "Taste is hardwired because you don't want the newborn to have to learn anything before it avoids certain problems that could kill it. Well, the first thing that's going to happen to a newborn is that if it doesn't eat, it's going to die. So you hardwire sweet to be very good, and you make mother's milk sweet," says Linda Bartoshuk.

Human breast milk derives 40 percent of its calories from lactose, otherwise known as milk sugar. Put a bit of sugar on your finger and put it in a newborn's mouth and you'll see that the baby will communicate to you nonverbally that *this is a good thing! Give me more!* Wiring kids with an innate love of sweet tastes helps ensure that they'll be able to nourish themselves.

Humans also need salt to survive, but babies don't have a preference for salt from day one, because they simply cannot taste it. Human salt receptors aren't mature at birth, even for the offspring of mothers who had morning sickness. Their saltaholic cravings, like all human babies', don't kick in until a few months after birth. There's a horrifying case of a hospital in suburban Sydney, Australia, that proved this fact in the worst way. An employee in the children's ward accidentally used salt instead of sugar to make the infant formula, which in the late 1960s required measuring and mixing. The newborn babies drank the formula without fuss, given their immature, nonfunctioning salt receptors.

Soon they sickened, but by the time they were diagnosed with salt poisoning, the mix-up had caused the deaths of four newborns. Adult humans also can die from salt overdoses, so imagine how sensitive a baby's system is to the effects of high levels of salt. Breast-feeding infants is safest for them, although most baby formulas these days already have the sugar included. If yours does not, move the salt container away from where you mix it up.

We know from research on salt preferences that infants develop a liking for salt at about four months of age. And then something interesting happens. Children under the age of three will drink salt water that a child over the age of three will reject. It seems that as the child's salt receptors mature, she becomes more in tune with the context in which the saltiness occurs. After three years of age, kids will reject salty water yet accept salty soup broth at the same level of saltiness. Most adults reject salt water, which just doesn't make culinary sense to us. Salty soup broth makes sense. Three years of age seems to be when food form and context become important in making decisions about what to eat. It's when we begin to make sense of our food world.

Babies universally reject bitter foods. There's a good evolutionary reason for this, too. Human babies are completely helpless when they're born and rely almost entirely on adults for their survival. If a baby were left on his own, how would he know what to eat? Most babies will put anything into their mouth, and most babies take immediately to sucking on a nipple, no instruction required. They have the instinct to get things into their mouth. But if babies—and children—would universally swallow anything, they'd be at great risk because they have undeveloped immune and digestive systems. An instinctive rejection of bitter foods is thus protective of human babies' lives. This explains why kids don't (usually) drink coffee, tea, or enjoy Brussels sprouts or broccoli. It's the remnants of their protective wiring hanging on. I asked Gary Beauchamp, director of Monell, if he had any advice for mothers who want their kids to eat more vegetables. His response was not to force children to eat anything. "Kids are smart not to like vegetables. They're onto something. You're fighting some real biology there," he said.

Certain kids don't like vegetables, and they're not just trying to make your life difficult. It has to do with their developing systems' protective mechanisms. Many vegetables have a bitter taste component, and rejecting bitter ensures that they won't poison themselves before they know better. This inborn hatred takes years to dissipate. Sometimes it mellows into a mere dislike of vegetables; sometimes it disappears altogether. This will depend on individual anatomy and genetics, both of which influence taster type. But mostly, it will depend on what children learn from their parents.

In-Uterometer

Cheese aficionados know that the flavor of a cheese depends on the flavor of the milk that was used to make it. For example, cheese made from cow's milk savors differently from cheese made from sheep or goat milk. Furthermore, the flavor of the milk depends on the food that the cow (or sheep or goat) ate. I vividly remember eating a piece of cheese in Cologne, Germany, because the sulfurous, green, vegetal flavor was so strong that I had to look down at my plate to make sure that I'd actually eaten cheese and not a vegetable. I thought I tasted notes of cauliflower or broccoli, yet the mouthfeel told me it was cheese. The green, vegetal flavors of the cow's diet had solidified into a medium-soft cheese. The experience opened my eyes to the potential complexities of cheese. In the United States, our cheese is made from pasteurized milk and, as with any food, when milk is heated—which is what pasteurization does—the volatile top notes of flavors are flashed off. In Europe, cheese is made from unpasteurized milk, and the resulting range of flavors takes you back to the field, the farm, or the pasture where it originated. You can literally taste the type of grass the cow grazed on.

With this as common knowledge, Monell Chemical Senses Center researcher Julie Mennella wondered whether flavor preferences could also be transferred from human milk to breast-fed babies. She hypothesized that Italian women gave birth to babies who were more tolerant of the flavors in the Italian diet, such as garlic and tomatoes, because their mothers ate these foods when pregnant, and they made their way to the developing babies in utero and after. Similarly, she wondered if Japanese women reared babies who were predisposed to like fish, and so on. Menella recruited pregnant women to participate in an experiment to see if this could be true. She divided the mothers-to-be into three groups. She fed the first group carrot juice during the last trimester of their pregnancies. The second group ate carrots regularly while breast-feeding their newborns, and the third group avoided carrots altogether.

Months later, Mennella brought the babies and their mothers back into the lab and had the mothers feed their babies plain and carrot-flavored cereal. Sure enough, the babies whose mothers had consumed carrots while pregnant or lactating liked the carrot cereal better than those born of mothers who hadn't eaten a single carrot while pregnant or nursing. They also made fewer of those adorable newborn *I can't talk but I'm gonna show you I don't like this* faces.

Mennella's research proved that exposure to flavors as early as in the amniotic fluid and in breast milk can influence babies' preferences. Her research was done with carrots, a fairly mild food, but imagine the results she might have gotten if she had done the research with garlic or fish. Whatever a pregnant or lactating woman eats, her baby will be exposed to, for better or worse.

I talked with a woman at the Senior Friendship Center in Sarasota, Florida, who grew up hating liver. She was forced to eat it as a child and never developed a liking for it. The first time she got pregnant, she avoided liver like the plague. Even the smell of it made her feel sick. This was also true during her second and third pregnancies. Then something happened after her third child was born: she miraculously started to gain an appreciation for liver. Of course (in what may be a cosmic case of payback), none of her kids would eat liver with her, just as she refused to do with her mother when she was a kid. When she got pregnant the fourth time, she ate liver while carrying the baby and while breast-feeding. Years later she would make liver for herself and her youngest. The only one of her children who had been exposed to liver in utero ended up being the only one who would eat it.

In order to raise kids who eat more healthfully later in life, mothers need to eat healthful foods themselves. If mothers want their kids to eat broccoli, they need to eat it themselves when pregnant and nursing. The same holds true for garlic, fish, and just about everything. Eating a varied diet of beneficial foods during pregnancy and lactation is good advice for the mother's health and for arming a child with a built-in preference for those same healthful foods. Think of eating salmon, cauliflower, and Brussels sprouts while pregnant as giving your developing baby a vaccine to ward off future vegetable rejection.

Aroma Diploma

We know that even days-old humans can detect smells. If you put something pungent under their noses, you'll definitely get a reaction. Their olfactory systems work from the get-go. But do smells mean anything to newborns, or do they have to be taught what smells mean? This question generates a lot of interest. As an example, let's choose a strong smell that almost all adults would classify as offensive: the odor of feces. A baby with a full diaper can play joyfully with zero care that it's stinking up the room. At what point in the baby's life does the smell of poop become negative? Does this rejection of fecal smells develop as the child ages? Or does the child learn that the poopy smell is bad by watching the reaction on the faces of his siblings, parents, and friends when he fills his diaper?

Most scientists believe that aroma preferences are learned, not innate. They argue that taste is constant around the globe, because there are only five tastes and these are inherent in a huge array of foods. Wherever a baby is born, his environment will contain bitter poisons that could kill him, and sweet substances that could nourish him.

Yet there isn't a single volatile aroma that exists in nature that—at a normal level—could kill or nourish an infant. Aroma preferences don't ensure the survival of an infant. But taste preferences and rejections do. Smells also vary wildly according to geography. It's unlikely that a baby in Scandinavia will smell corn tortillas cooking on a griddle. And it's unlikely that a baby in Mexico will smell the rotten fish surströmming. If Mexican babies were born with a preference for—or more likely against—surströmming, they would never even know it because they'd never run across it (or run away from it). This would be extraordinarily wasteful, cluttering our brain with preferences for or rejections of aromas that the vast majority of humans will never experience.

Even though our scent preferences are acquired through our culture, they still have the power to define what we consider delicious or disgusting, to help turn appetites on and off, and to incite emotions. The bottom line is that we learn to like or hate aromas, unlike tastes, which we're born loving (sweet) or hating (bitter).

Food Phobia

If you have a child who won't eat something, count your blessings. This indicates that he has a healthy fear of the unknown. This is the same fear that will stop your kid from jumping off a roof, taking a ride with a stranger, or petting a wild animal. We are naturally suspicious of things that are new to us, a fear called *neophobia*. At the dinner table, neophobia can be a source of great tension and frustration.

Luckily for the survival of our species, parents usually protect their children from eating spoiled, poisonous, or otherwise harmful food. But once a young child is out from under his parents' constant mealtime oversight, between the ages of about two and three, neophobia kicks in. Fear of new foods declines with age. Twentysomethings accept more new foods than kids in high school, who accept more new foods than those in junior high school, who accept more new foods than children in elementary school, and so on down to about the age of two. Under the age of two, children will try (but not necessarily eat) just about anything offered to them.

Experts estimate that most children need five to ten exposures to a new food before accepting it. This doesn't mean they'll be clamoring for more lima beans after their sixth try, but they'll be more likely to start to accept the food. The problem is that most parents get frustrated after three or four tries. It's best to let some time elapse between trial tastings of the new food. It takes patience and desire to get your kids to eat more healthfully. A week or two later, it's time to try again.

Don't use dirty tricks to get your kids to eat their lima beans—they will backfire. If you promise your kids a deep-fried, chocolate-dipped goo-goo cake to reward them for doing their homework or chores, you are unconsciously instilling in them a desire for deep-fried, chocolate-dipped, goo-goo cake. If you bribe them to eat their lima beans with the promise of an extra hour of online gaming or a trip to the toy store, you're stigmatizing beans as bad foods and ensuring that they'll like those lima beans less. If you want to instill healthy food attitudes, teach your kids that there are no bad foods. Even a deep-fried, chocolate-dipped, goo-goo cake would be a good food to teach them about taste concepts such as "unbalanced sweetness" and "overly fatty mouthfeel."

Food neophobia is protective, but these days we rarely encounter (or need to eat) potentially dangerous foods. Unless you're eating blowfish sushi or high meat, you're unlikely to come across much poisonous or spoiled food in your current environment. Yet food neophobia exists for one very real reason: In developed nations people are afraid of new foods because they think they're going to taste bad.

People are more likely to eat something new if they are in the right state of mind. One state that you want to steer clear of is arousal—or, in other words, activity and excitement. If you want your kids to try something new, don't serve it to them at a party or at holiday time. New foods—like most new things in general—are inherently stimulating, so the combination of a novel food and a novel situation can push some kids over the edge. On the other hand, adults seek out new foods as a form of stimulation and entertainment, and some television shows make a huge fuss when the host eats something new for the first time. This is the wrong type of environment to enable the eater to enjoy a new food, says Patricia Pliner, who studies eating behavior at the University of Toronto, but it comes back to individual preferences. Some people are thrill seekers when it comes to food and some are not.

If you want people to accept something new, give them a lot of positive information about it, which helps match their expectations to the reality of what they're going to eat. The first time I was presented with foie gras, which I was told was goose liver, I was horrified because I despise the flavor and texture of liver. If someone had told me that seared quickly, salted perfectly, and served with toast points, it tasted like my Grandma Ruth's roast chicken gravy—fatty, salty, chickeny, and loaded with umami—I would have been more likely to try it. Telling people that pomegranate juice was healthful certainly made them drink more of it than they did before they had that information.

Repeated exposure to a food helps adults and children accept it. I used to hate everything about canned tuna fish. Then I took on a tuna project at work. Having to taste canned tuna over and over and over instilled in me a level of familiarity with the product that the smell had kept me from developing. It took me about a year of intermittent tastings to gain appreciation for and discrimination between types of tuna. This allowed me to discover albacore tuna that's packed fresh in oil—a completely different sensory experience: savory, complex, unctuous, and absolutely satisfying for a healthful meal. Much of the canned tuna sold in the United States is cooked first, then cut or flaked off the bones and put into cans, where it undergoes a second cooking. If you want the best-tasting tuna, look for the fresh-packed kind that is cooked only once. This means it's packed raw before being cooked during the canning process. As you can see, with repeated exposure I became quite the tuna connoisseur. So can you with any type of food.

In a familiar situation, people are also more likely to try new foods. Imagine someone asking you if you'd like to try a new kind of meat called *nutria*. If you were at your best friend's house, you'd be more likely to give it a try than if you were in the break room at work where that guy from accounting told you he'd shot it over the weekend and offered you a bite of his nutria sandwich.

One of the best ways to get someone to try a new food is to employ the concept of *flavor principles*. In the early 1980s, Elisabeth Rozin wrote about the "recurrent flavor combinations and cooking techniques" that give an ethnic food its characteristic signature. For example, Chinese food usually contains ginger, garlic, and scallions. It is often stir-fried or steamed. These recurring flavor principles and cooking techniques make Chinese cuisine different from Japanese and Italian. Rozin believed that flavor principles were useful for introducing new foods within a culture. For example, if you wanted someone who was raised in China to eat a new type of meat, such as nutria, you could stir-fry it with ginger, garlic, and scallions to make it more tempting because the preparation and flavors would be familiar.

Patricia Pliner tested this theory in 1999 at the University of Toronto. Her team offered taste testers a novel food (such as *parval*, an Indian vegetable, or *gathiya*, an Indian snack chip) either plain or with a familiar sauce. When the novel food was paired with a familiar sauce, people were more likely to try it. When you are trying to get your children to eat a novel food, sometimes all they need is a condiment (ranch dressing or ketchup) served alongside it. Better yet would be to cook it in your family's own flavor principle. In my household that would mean brushing it with fresh extra virgin olive oil, grilling it over high heat, then seasoning it liberally with sea salt and a generous squeeze of juice from fresh lemon wedges. In fact, I'll bet I could get Roger to eat nutria if I prepared it this way.

Pliner's next experiment required hiring "confederates" to act like test subjects, when in reality they were members of her research team. Each confederate was paired with an unsuspecting tester and the pairs were told to choose from a list of familiar, safe foods or unfamiliar, novel foods. The real test subjects were more likely to try new foods when they were paired with a confederate who modeled the behavior of choosing novel things—like nutria—over more familiar foods, like chicken or beef. You are more likely to do something if you see someone else doing it.

Nutria, by the way, are large semiaquatic rodents that were imported into Louisiana from South America for the fur-farming industry. They escaped (or were released) into the wild and have become destructive pests. In an effort to control the nutria population, the state of Louisiana tried to promote them as a source of protein. The state got the word out and put up websites with regional recipes to encourage capture and consumption of the critters, including Heart-Healthy Crock-Pot Nutria, Nutria Sausage Jambalaya, and Nutria Andouille Sausage Gumbo. Yet even appropriate use of flavor principles couldn't get the Louisiana public to eat wild rodents. Imagine New York taking this approach to deal with its rat problem. *Rat Reuben on Rye* does have a certain ring to it, though.

That Time of the Month

There is one day of the month when no matter how much salt I use, my food tastes bland. On this day I do not make any decisions in the lab at work. Instead I rely on a team of other people. When I cook dinner, Roger knows what's up by simply watching me twisting the salt grinder as if I'm trying to wring it dry. Every woman has experienced something similar, whether it's a craving, an aversion to a certain food, or a smell that simply nauseates her for no other reason than "It's that time of the month."

The most intensive study to date on women's sensitivity to aromas throughout their menstrual cycle ran up against the fact that women have different cycle lengths with different hormones spiking at different times. After the authors normalized the data, they found a peak in the ability to detect smells at midcycle: in other words, around the time of ovulation. This was true of women both on and off birth control pills. They also found that olfactory sensitivity practically mirrors body temperature when plotted on a graph.

Women are in general both better tasters and better smellers than men, as has been proved in numerous studies in which women usually outscore men. Monell's Johan Lundström says that this is due not to their superior anatomy, but to the fact that women pay better attention to the task at hand. But not all women are better tasters than all men. In your household, there might be a male HyperTaster who is much more sensitive to tastes than a female Tolerant Taster.

Smell That Bun in the Oven?

Everyone has a story of a wife, a friend, a sister, or a colleague who became so sensitive to odors when she was pregnant that she could no longer tolerate the offending smell at all. Sometimes, a woman even becomes intolerant of her spouse's smell, which can be a harbinger of incompatibility and an impending split.

In one study, 67 percent of pregnant women said they'd experienced an increase in their sensitivity to aromas at least once during their pregnancy. Self-reported data are notoriously unreliable, and other studies that have tried to scientifically prove this finding have been inconclusive.

Whether or not pregnant women can smell more acutely than those who aren't pregnant, you would think that they'd be able to know what it is they're smelling. After all, they are the gatekeepers for their developing babies. If they can't smell the difference between noxious fumes and baking bread, that poor child is in danger. But to date, no study has found that pregnant women show an overall, general increase in their ability to identify odors. In fact, in two different studies, pregnant women were on average less likely—or showed the same ability—to identify smells as did nonpregnant women.

Seventy-five percent of women report that some smells were less pleasant when they were pregnant. Some studies suggest that the unpleasantness—or disgustingness—of odors is strongest in the first trimester. This makes good sense, since a pregnant woman's level of immunity drops during this period, making her more susceptible to toxins, illness, and disease. Bitter tastes are most intense and liked least during the first trimester. If a pregnant woman's digust-o-meter is very sensitive, she'll be less likely to eat food that is dangerous to the baby. The problem is that we have very little data on women in their first trimester. It's hard to recruit women to be part of an aroma study in their first 90 days since many of them don't know they're pregnant until 45 to 60 days in.

In the second and third trimesters, bitter tastes are tolerated a little bit more (or hated a little bit less), as are the sour and salt Basic Tastes, and this change encourages women to ingest a more varied (read: healthful) diet.

Throughout pregnancy, however, a woman's reaction to aroma is heavily dependent on what she smells and is related to her own likes and dislikes.

The Taste of Maturity

One of the cruelest things about growing old is that our ability to smell almost inevitably degrades. Half of people between the ages of sixty-five and eighty lose some of their smell functioning. And more than 80 percent of those over eighty years of age have a compromised sense of smell.

I spoke to a group at lunch one day at the Senior Friendship Center in Sarasota, part of a network of nonprofit facilities that provide services to those over fifty years of age, because I wanted to talk with older adults who were experiencing this loss. One woman just about broke my heart. She was slowly losing her passion for eating, which was devastating to her.

"What else do we have left?" she said wistfully of seniors like her who have started to lose their enjoyment of food. "We can't drink because we're on medication. We can't dance because our bones have become brittle. We love to eat. And when that goes, it's depressing. The only reason I eat is because I have to eat. And that scares me."

Then she reached into her purse and pulled out a tangerine. "This I can taste," she said, gripping the piece of fruit as if her life depended on it. Well then, use it, I told her. I asked her if she was enjoying her chicken with black beans and rice. She wasn't. "I'm just chewing," she said. So I told her that the next time she came for lunch, she should ask the cooks to cut a tangerine into wedges for her to squeeze over her meal. If something works for you, why not embrace it? That's the takeaway. Figure out what works for you. Explore. Crank it up. Don't just sit back and accept the loss.

The Taste and Smell of Alzheimer's

Jennifer Stamps, a graduate student who works for Linda Bartoshuk at the University of Florida, studies olfaction and the brain. A few years ago, she was living next door to an elderly man who was in the early stages of Alzheimer's disease. Stamps brought home the UPSIT, the University of Pennsylvania Smell Identification Test, one day and administered it to her neighbor. He scored 25 out of 40, a typical score for an Alzheimer's patient who has started to lose his sense of smell.

About six months later, the neighbor told Jennifer that something was wrong. He could smell food while it was cooking, but once it was in his mouth, food just didn't taste right. Everything "tasted" like salty cardboard. Jennifer pulled out the UPSIT again and retested him. Same score: 25 out of 40. On the surface, nothing had changed. But the UPSIT is a scratch-and-sniff test, meaning you don't put anything in your mouth. It measures only nose-smelling. The neighbor had barely complained about his loss of nose-smelling, probably because it had occurred gradually, but over the past six months something different had happened. He had lost his ability to mouth-smell.

"When he lost his retronasal (mouth-smelling), it was very dramatic for him. Very distressing. He lost thirty pounds in three months as soon as this happened," said Stamps. "It had a much greater impact on him, on his health, on his well-being, pleasure for life. Everything. He got very depressed, lost a lot of weight. His cognition declined even more, and very dramatically, and his health just went down the tubes."

Stamps was flummoxed. She asked Bartoshuk what she thought was happening. Why hadn't the test captured this type of olfactory loss? Was it even possible to capture this?

"You haven't caught it because it's *taste*, Jennifer," Stamps remembers Bartoshuk saying to her, meaning that she had missed this loss because she had been testing the functionality of the wrong sense.

Stamps tested her neighbor next time for taste, not smell, and found that the back of his tongue was completely useless for tasting; this meant that the glossopharyngeal taste nerve was completely dead. Also gone were the areas of the tongue that connect with the left side of the chorda tympani facial nerve. His ability to mouth-smell disappeared when he lost some of his sense of taste. This is the same phenomenon that happened to Bartoshuk's patient who had slashed her taste nerve by licking the inside of a metal can.

Stamps began to test people who had experienced partial flavor loss like her neighbor, asking them to rate the intensity of various foods by both nose-smelling and mouth-smelling. Foods with high levels of chemesthesis, or tactile burn, such as curry, mustard, vinegar, and garlic, seemed to be the most resistant to loss. Foods without a tactile component (such as grapes, butter, and apples) were most susceptible to loss.

Stamps is now using this knowledge to test whether adding an Irritaste to foods will help those with taste loss get more aroma from their foods. Her secret ingredient is cayenne pepper. She hypothesizes that adding the tactile burn of capsaicin at a level below which it is detectable may increase mouth-smelling and, therefore, overall flavor perception and ultimately the enjoyment of food. The cayenne stimulates the touch nerve that gives the olfactory system the input it's not getting from the taste nerves.

A woman with Parkinson's disease visited the University of Florida's memory disorder clinic complaining of flavor loss. Stamps gave the woman a full taste and smell work-up, which showed dramatic smell loss. Then Stamps gave the patient her untested "cure," dosing grape jelly with increasing levels of cayenne pepper, all of them at a level that was below what the patient could detect.

To set the baseline, Stamps served the patient plain grape jelly. The woman said she got a musty flavor, but that was it. When she tasted the first, lowest concentration level of cayenne-and-grape-jelly concoction, the patient said the same thing. Musty, but that's about it. At the next higher concentration, she looked straight at Stamps and yelled, "Grape! I got grape!" as she experienced exactly what Stamps had hoped. The right amount of tactile stimulation (from cayenne pepper) had kicked this woman's flavor perception into gear.

"We don't know how or why—if it's tricking the trigeminal nerve into carrying the olfactory information, or what," said Stamps. "I have no idea how it's working. But it is."

The Parkinson's patient got teary-eyed and emotional upon experiencing the simplicity of grape jelly again for the first time. She had tasted a glimmer of hope in what she thought would be a depressing, downward spiral into flavorlessness. Stamps gave the patient a list of Irritastes and told her to "go play" at home in her kitchen. The woman left Stamps's office as giddy as a child on her way home from the toy store.

Dentition Condition

Oral health can also affect older people's ability to taste. When you age, you produce less saliva than you do when young, so you're less able to moisten foods. Eating crunchy or dry foods—pretzels, rice cakes, and croutons—is less appealing. Of course, in order to even bite into a pretzel without effort or pain, you need a healthy set of teeth and gums or you'll have difficulty chewing. Without chewing, you can't release the volatile aromas from food. Chewing gives you more of everything the more you chew it: basic tastes, aromas, textures. The University of Connecticut's Valerie Duffy has found that older women with dentures had more complaints that they were not experiencing food fully than those with their own healthy teeth and gums.

According to Richard Doty, director of the Smell and Taste Center at the University of Pennsylvania Medical Center, people who already are genetically prone to Alzheimer's disease turn out to have a ninefold higher risk when they also report problems with their sense of smell. Also, substances from the environment can enter through the nose and reach the brain, triggering diseases such as Alzheimer's, Parkinson's and Creutzfeldt-Jakob's. The best way to avoid this type of nasal exposure is to avoid the toxins that cause harm in the first place. Doty recommends avoiding herbicides, pesticides, and heavy metals, and getting treatment for infections immediately. This is good advice in general, though we know very little about these devastating diseases.

Use It or Lose It

If you or a loved one suffers from a diminished sense of smell or taste, there are a few things you can do. It's likely that someone with diminished senses has already started salting his food more heavily or adding more sugar to coffee—the first, most logical reactions. If you want to get more flavor from your food, you add salt. The problem, though, is that most older adults, especially those with hypertension or diabetes, don't need more sodium or calories.

Taste is much more resilient to the effects of aging than smell. It's likely that the sense of taste is fine, so they'd be better off squeezing tangerines or lemons over their food, as I told the woman in Sarasota to do, than adding more of the Basic Tastes. What they are really craving are aromas, which citrus fruits have in spades. Using aromatic citrus probably won't harm an older person's health and won't add calories, sodium, sugar, or guilt. Buy fresh lemons and serve a wedge with everything from already-dressed salad to subtle foods such as potatoes or rice. You can also start playing around with cayenne or other hot peppers. Be cautious, though, as many hot sauces contain sodium, although when used sparingly, the level tends to be minimal. Try smoked peppers, which add almost no sodium or calories. You can probably find ground chipotle chiles at your grocery store and a multitude of other dried chile powders at Hispanic markets.

University of Florida's Jennifer Stamps wants to do research on professional chefs in the future. She has a theory that they do not get dementia as frequently as others because they're constantly using their senses of taste and smell. She thinks that rigorous use of their olfactory sense, in particular, keeps their brain in shape, like a muscle: if you don't use it you lose it. She also believes that people who eat the same thing every day, like her grandfather, aren't building new olfactory cells. By eating more varied, new foods, you challenge your brain, keeping it young. Her grandfather died of a neurological disease and she's convinced there's a connection. If eating a varied diet with thrill-seeking abandon is a way to stave off senility, then I say bring on the nutria.

Part Two
The Basic Tastes

8

Salt

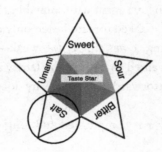

It's tomato season as I write this, my favorite time of year. There are only a few months of this ecstasy, when the tomatoes are pregnant with juice, thick-skinned, and calling to me from every grocery store, farmers' market, and restaurant menu. My father grew tomatoes in our backyard, and we ate them every night during the summer. On Sunday he'd fry slices of them with a crispy breadcrumb coating. They are one of the flavors of my childhood.

For forty-some years, I've been salting my fresh tomatoes. In my youth, I sprinkled Morton's iodized salt on them, as my dad taught me to do. Later, I added cracked black pepper. I went through a brief phase of dousing them with balsamic vinegar. In the early 2000s I discovered sea salts in all hues, crystal sizes, and flavors. In the past year I've been treating myself to a drizzle of tree-fresh olive oil from our friends' grove and some freshly ground salt, my current favorite being Himalayan Red Mountain.

But this year I decided I was going to go through tomato season without salting my tomatoes. I had become so accustomed to salted tomatoes that I thought of them as a single flavor, fusing them in my brain as I do buttered popcorn or milky chai tea. I never ate one without the other, so I never really gave much thought to their individual flavors.

Eating an heirloom tomato without salt was like getting intimate with a former lover for the first time in years. I knew the curves and contours of my tomato, but I never realized just how beautiful it was naked. Its forgotten perfume aroused my sense of smell: a whiff of earth, a sniff of vegetal, and a dash of musty vine. The height-of-the-season sweetness made the juice softer, less acidic. And the ripe, red flesh was full of free glutamates, which mean savory umami goodness. I was in love again with something I eat almost every day.

Yet my affair with the naked tomato lasted only about three weeks, after which I returned to salting. I was simply too weak—or is it that salt is simply too good? I really wanted to believe the latter, so I went to the experts with one simple, straightforward question. Why does salt make food taste so much better?

As always, the answer is complicated and long—unlike tomato season, alas.

What Is Salt?

We use the word *salty* to refer to the taste of many foods. But the prototypical pure salt Basic Taste is the compound called sodium chloride. This is the stuff we know of as table salt.

Paul Breslin of Rutgers University has spent a lot of time studying salt. According to Breslin, the sodium part of sodium chloride is what makes salt taste salty; the chloride part of it "enables the sodium to do its thing."

What that *thing* is, though, is still a mystery. "We don't know how salt taste works," Breslin says, meaning that we don't fully understand how it works at the microscopic level of the taste receptors within our taste buds. Scientists are working to identify the salt receptor. Nonetheless we do know how salt works to make things taste better.

The Buddy System

The short scientific reason why I caved in so easily is that salt actually makes savory foods such as tomatoes taste better.

"Something that is purely salty and something that is purely umami [savory] won't taste nearly as desirable as the combination of the two together," says Breslin. Tomatoes are high in the umami Basic Taste, so the sensory input from the combination of savory tomato flavor and salt is greater than the sum of its individual parts. This has been proved for many different foods. We consider a chicken soup more chicken-y if it contains salt than if it does not. The challenge in cooking and seasoning food is to heighten the savory flavor of the chicken without going overboard on the salt. Chef Joshua Skenes of the restaurant Saison in San Francisco perfectly captures the challenge in using salt to bring out other flavors.

"We look at salt not as something that you can just throw on food to make it taste good, but as something that pulls the flavors and extracts the flavors from food. You don't want to taste salt," he says, "you want to taste the ingredients. You want to salt the food so that you can taste the most natural purity of the flavor in the food to the fullest possible extent but *not taste the salt*."

This was exactly what I was doing by salting my tomatoes. I was not looking to taste the Basic Taste salt. I was simply looking for more tomato flavor, which was partly achieved by salting, since that makes the umami Basic Taste more intense.

Salty as Salt

The simplest way salt works is that salt tastes salty and we innately crave this taste. The reason we crave salt is simple, and actually quite elegant. We have evolved to crave salt to ensure that we eat enough sodium to sustain life. A mineral that's found in many places in nature, including human cells, sodium is essential to regulating the water balance in cells, and plays a role in nerve and muscle function. Our bodies generally maintain the perfect amount of sodium in our blood, which is such a narrow range that you'd think a micromanaging accountant was checking on the numbers every day. If the sodium level in our body gets too high or too low, the kidneys and heart bring it back into range. This balancing act is done without a backup reserve of sodium for times when we might need it, and we are constantly losing salt through urine, feces, sweat, and tears, which (hopefully) you shed only from joy. Without a way to store excess sodium in our bodies, we have to make sure we get it from the foods we eat or drink. The result of losing too much sodium can be death.

A person who is severely dehydrated thirsts for water so desperately that he will drink whatever is available. People who are lost at sea drink seawater (which we dislike and which further dehydrates us) and urine (which disgusts us). But if you are deprived of salt, you will not crave salt in the same way a person with a life-threatening thirst demands, hallucinates about, and obsesses over water. Humans clearly have an appetite for salt, seeking it out in all types of food, but this is different from salt hunger, which could save our life by making us crave salt in situations when we need it. For some strange reason, we don't read our bodies well enough to know when we're dying from salt depletion.

Why do we not hunger for salt when the sodium balance in our body is out of whack? Your craving for salt is affected when you've lost sweat from exercise. Your sweat contains sodium, which means that salty foods and drinks will taste more palatable than before you exerted yourself. Even so, you won't spontaneously reach for the salt shaker to replace this loss. You will simply drink and eat food, some of it containing salt, some not, until your body is back in balance. Human survival depends more on getting water than it does on getting salt. So we crave water first and salt second.

The Role of Salt

Salt serves an important role in the glorious transformation of foods by cooking. When you cook certain foods, they brown on the outside, which can change and intensify their flavors. Browning also creates new flavor compounds that humans generally find desirable. Uncooked bread dough, for instance, is pale white and doesn't taste very good if you have the guts to eat it (or a job like mine). The aroma and golden crust of a perfectly baked sourdough baguette, on the other hand are crave-able. A raw steak doesn't hold much appeal (unless you doctor it up with the makings of tartare or carpaccio, both of which rely on the addition of other ingredients). But the browned, crisp edges of an expertly grilled steak can make your mouth water.

These types of browning are due to the Maillard reaction. Salt helps release the volatiles that occur during the Maillard transformation that makes food more appealing, such as the aromas of bread baking and steak grilling. Says Paul Breslin, "The smell of fresh-baked cookies, the smell of fresh-baked bread, is not the same in the absence of salt."

I can attest to this. On a trip to the Italian countryside in Tuscany, I ate some of the best food of my life. It was my first trip to Italy and I relished every bite. I had a tiny scoop of gelato every evening. I reveled in multiple plates of pappardelle with rabbit ragù and cemented my love affair with the Italian bubbly wine prosecco. Everything seemed to taste better than I expected, as if the Tuscan air had deposited a layer of deliciousness atop already amazing food. That is, everything except the bread.

Tuscan bread is made without salt. It is pale and bland and lacking in characteristic bready flavor. Theories abound as to why this region of Italy, alone, bakes bread without salt. One story is that ancient tariffs on salt upset the Tuscans so much they staged a tea party–like backlash: they simply stopped using it in their bread. I'm skeptical about this theory because the pastas and cured meats of Tuscany are wonderfully full-flavored and full-salted, although these give rise to another theory. This one holds that the salty meats and cheeses of Tuscany demanded saltless bread as a foil. I'm not buying this, either. Spain and France have long histories of curing meat and making cheese and their bread contains salt, which enhances the pairing. The rest of Italy eats salty meats, cheeses, and, yes, salty bread. Why in the world would this part of Italy hold on to this taste-killing tradition, regardless of its origin? Paul Breslin backs me up: "Bread with salt in it has more browning reactions occurring and it will smell more like the classical sort of fresh-baked bread aroma, which many find more desirable." Except, apparently, the Tuscans.

Mutual Suppression

My grandfather used to salt his grapefruit half, which my grandmother served with a maraschino cherry in the middle. I always wondered: why the salt? He said that the salt made the citrus fruit taste sweeter. As a child, I thought, How could salt—which tastes salty—make grapefruit taste sweeter? It's the same grapefruit half before and after salting, and he was adding salt, not sugar. But when I tasted it, I found he was right.

My grandfather knew something from experience that has only recently been explained by science, an effect called mutual suppression. The right amount of salt makes grapefruit taste sweeter. The right amount being a threshold level, which doesn't make the grapefruit taste salty because it's just below your threshold of detection.

In fact, lemonade is another example of mutual suppression, a sort of canceling-out phenomenon whereby the Basic Tastes sour and sweet suppress each other. Imagine three batches of liquid: the first is two quarts of pure unsweetened lemon juice; the second is two quarts of sugar water; the third is lemonade made from two quarts of lemon juice plus two quarts of sugar water. If you tasted each, you would say that the pure lemon juice was very sour and the sugar water was very sweet. You would find the lemonade less sweet than the sugar water and less sour than the lemon juice.

When you add salt, though, something curious happens. Salt acts as something of a taste superhero, thwarting the bad guys while assisting the good guys. When you add salt to food, it suppresses "bad" tastes, such as bitter or sour. But salt isn't as punitive to the "good" tastes of sweet and savory. Salt releases the desirable flavors from suppression by the bitter or sour tastes like Superman freeing Lois Lane from Lex Luthor. The result of this salt superheroism is that a pinch of salt can make bad things taste less bad and good things taste better. If you perform the Experiencing Mutual Suppression exercise at the end of this chapter you'll taste this phenomenon firsthand. Sweetened bitter tea tastes less bitter and more sweet when you add a touch of salt. My grandfather *was* making his grapefruit taste sweeter with salt by suppressing the sour and bitter tastes while the sweetness actually stayed the same. After adding salt he could taste the sweetness more clearly because it was released.

We often use this knowledge of taste suppression at Mattson. We may add a tiny dash of salt to a formula where you would not expect it, such as hot chocolate or dessert sauce. Chefs also use this technique, many without knowledge of the underlying science. They just know that the end result is that a recipe tastes better with a tiny bit of salt added to it.

Salting Out Volatiles

The chapter on smell explains that the aroma of a food comes from its volatile compounds. In other words, as a soup simmers on the stove or a pie bakes in the oven, the volatile ingredients in the soup and pie start to waft off. The compounds in the garlic, onion, and celery or the apples, butter, and cinnamon move from the cells of the food and into the air. That's when you start to smell the wonderful aroma of Grandma's matzoh ball soup or Auntie's apple pie. Adding salt to food makes it release more aroma—the salt nudges aromatic compounds out of the cells of the food so that they volatize and you can smell them. Salting raw tomatoes makes them *smell* more tomatoey. And since smell makes up much of flavor, salt increases the signature flavors of a food.

Salt in Processed Food

We perceive salt through a receptor channel in the taste bud, whereas we perceive sugar and bitter through a hand-in-glove type of receptor connection that is easier to fake. Hence the proliferation of sugar substitutes on the market, while the quest for the holy grail of the food industry continues: a salt substitute that tastes like salt.

Salt is often used in processed foods to assist in the "functionality" of the food—salt makes meat seem juicier when it's frozen and recooked at home in your skillet, on the grill at a restaurant, or in your lunchtime frozen entree. This happens by osmosis: water moves from a salty marinade into the cells of the meat so that there's more juice in the meat when it starts cooking. Since liquid escapes in the cooking process, using a salty marinade results in juicier meat. Salt is also used as a preservative in many foods, such as deli meats, hot dogs, and soy sauce, controlling the growth of harmful bacteria. And of course food manufacturers add salt to make food taste better.

Without salt, many food products would be unrecognizable. This was dramatically evident when a *New York Times* writer attended a rather unusual tasting at Kellogg's laboratories in Battle Creek, Michigan:

> As a demonstration, Kellogg prepared some of its biggest sellers with most of the salt removed. The Cheez-It fell apart in surprising ways. The golden yellow hue faded. The crackers became sticky when chewed, and the mash packed onto the teeth. The taste was not merely bland but medicinal . . . They moved on to corn flakes. Without salt the cereal tasted metallic.

Salt clearly serves a superhero role in some packaged foods, but the food industry's reliance on salt goes way beyond functionality. In some cases, the overreliance on salt is unwarranted. Adding salt makes product formulation easier, even in products where salt isn't used for its preservative effect. Salt is also cheap, so it's a way to add more flavor without having to add more of the expensive ingredients that give food its characteristic flavor, such as meat, vegetables, cheeses, or herbs.

Let's take canned and frozen foods as an example. There's absolutely no preservative role that salt plays in canned foods. It is there only to make them taste better. The same is true of most frozen foods, with the exception of the moisture-holding property proteins like chicken and beef mentioned earlier. Canning and freezing eliminate most of the need to use salt as a preservative. The salt problem in the food industry is a "catch-22" at this point. Americans are used to very salty food, so the industry has to deliver upon this expectation or consumers won't buy their products. The result is that we've all just gotten comfortable in this salty laziness. Salt doesn't require you to think too much about what you're eating. Salt is salty and that tastes good. Now, on to the next bite.

Salt and Health

The evolutionary elegance of our craving for salt has become a public health issue in industrialized countries where people don't sweat enough—due to lack of physical exertion—but still crave salt as our caveman ancestors did. The result is that we eat far more of it than we lose. Overconsumption of salt can lead to high blood pressure and other health problems.

If you eat a lot of salty foods, the sodium from the salt enters your bloodstream, pushing your fluids to the edge of normal function. To even things out, your body begins dumping water into the bloodstream, trying to maintain that narrow range of sodium. This extra volume swells your veins and arteries, making everything move a little faster. The result is that your blood pressure goes up.

Americans now eat far more salt than they used to, and more Americans have high blood pressure. The idea that the former trend is causing the latter makes so much intuitive sense that many scientists have accepted it, even though large-scale trials and epidemiological studies that attempt to link high salt intake to high blood pressure have shown mixed results. Any epidemiological study, in which you are looking at a disease in a lot of people, is complicated by all the ways that each of those people is different—this one eats lots of salt, but sweats it out all day; another eats lots of salt but doesn't sweat. In addition, controlling our blood pressure is not a simple mechanism with salt on one side and water on the other. It's more like a system of dams, canals, and pipes that controls the movement of snowmelt from the mountains, through farmland, into cities, and finally to your tap. Blood pressure depends not just on sodium, but also on potassium (which constricts or relaxes blood vessels), calcium, sugars, and hormones.

Jessica Goldman was forced to learn about this and a lot more than she ever wanted to about sodium intake when she was in her twenties and a series of health crises led her to eliminate salt from her diet. She was raised in Palo Alto, California, in a family where dinner meant take-out much more often than home-cooked food.

"No one cooked. Everything came from a take-out box. Chinese food, pizza, Japanese, those were our favorites," says Goldman about the food behaviors in her childhood home. The family members loved to eat; they just didn't love to cook. They were food- and flavor-focused to the exclusion of health.

"We were the family that took out the salt shaker and dumped it on our food before we had even tasted it," she said. Her favorites were fried chicken, French fries, and macaroni and cheese. Salt, salty, and saltier.

During her junior year abroad in Italy, she was diagnosed with celiac disease, a genetic disorder also known as gluten intolerance. Gluten is one of the main components in wheat and wheat flour. As a result, she had to severely limit what she ate, in a country renowned for its (wheat) pasta, (wheat) bread, and (wheat-crusted) pizza.

"It was horrible," she says about subsisting on *salumi* and cheese. Even more horribly, when she returned home she found out that she didn't even have celiac disease and that she had missed out on *pizza bianca* and pasta *primi* for nothing. Yet that period was early training for how to eat a severely restricted diet, which she would eventually need to do.

When she got back to California to start her senior year at Stanford, she arrived with an extra 40 pounds of weight on her normal 105-pound frame. She didn't even look like herself. This wasn't Parmesan and prosciutto weight, but excess fluid in her body. A week later she was also having seizures. Her bone marrow wasn't working. Her kidneys were failing. Her body systems were shutting down. She was told to put herself on a kidney organ transplant list and start life-sustaining dialysis. Eventually Goldman found out that she had a type of lupus that had attacked her kidneys and brain.

While she credits Western medicine with saving her life, she looked into the benefits of going on a renal-failure, or low-salt, diet to minimize her need for medications and dialysis. Was it possible, she wondered, to control her kidney disease by controlling the food she ate? She asked her health-care givers for advice and got a pamphlet from one doctor that made her laugh out loud with its basic but vague suggestions, such as "Don't eat soup." It didn't tell her what she *should* eat, so she decided to educate herself.

"I really made it my ultimate job to figure out how to keep myself healthy and off medications and treatments by regulating my diet, making it as strict as possible and giving my body as much room as possible to do as little work as possible," she said. She wanted to relieve her kidneys of the job of keeping her sodium level within the necessary narrow range.

Goldman had to go on a really, really low-salt diet, because her kidneys didn't work when she ate a normal diet. When she first removed every bit of added salt from her diet, she said, "It was definitely bland. When you don't taste salt, you think, *Oh God, there's no flavor in this.*" But then things began to change for her.

"As soon as my taste buds adjusted to not needing salt anymore or not expecting it, all of a sudden eating a red bell pepper was the most extraordinary thing. You really taste the natural flavors of food. It's been an unreal experience. I get to enjoy produce and protein for what they actually are."

Keep in mind that Goldman came from a family that didn't cook, even though she eventually had to learn how. Even more of a challenge was sustaining this diet when she was eating out—her default way of procuring a meal. The first type of restaurant she felt safe trying was a steak house, because she knew that she could find at least one type of meat on the menu that hadn't been marinated or seasoned. And there were lots of yummy side dishes that a steak house could do without adding salt, like baked potatoes and salad without the dressing. One night at a steak house she spoke with her waiter in excruciating detail about her dietary needs. He assured her that everything would be taken care of by the chef. When her steak was placed in front of her, she carved off a bite, and the very instant she chewed into it, she said, "Oh my god, it is sooooooo salty." She figured the chef had salted the steak and she was going to have to send it back.

The chef came out of the kitchen to talk to her. He looked her straight in the eye and said that he had cooked the steak himself, and he could assure her that there was no salt added. Then she realized what had happened. She had just tasted a piece of premium, aged beef—grilled perfectly over high heat—without salt for the first time in her life. What she had experienced was the pure, unadulterated flavor of the meat, not the seasoning that was applied to it. The meat, which naturally contained sodium, didn't need any added salt. It was that good.

"That was a real moment for me," she said. She had crossed over from mindlessly seasoning her food—an act that can obscure flavor in a frumpy muumuu of salt—to discovering the sexy, sensual, erotic flavor that's locked inside it.

These days Goldman is bolder in dining out. Her latest restaurant meal was at Frances, an acclaimed restaurant in San Francisco's Castro neighborhood. She has perfected her method for dining out by calling ahead to tell the restaurant staff about her condition, and instead of focusing on what she can't eat, she focuses on what she can. She asks to speak with the chef who will be cooking the night she'll be dining. She says that the chef can use unsalted butter, oils, herbs, vinegars, garlic, and anything else that doesn't contain salt. Frances served Goldman a piece of perfectly seared unsalted tuna in an unsalted tomato broth with cucumbers and jalapeños. Goldman said the dish's lack of salt "actually allowed the rest of the flavors to stand out. It was fantastic."

She has been on this extremely-low-sodium diet now for over six years and has learned a number of techniques to add flavor to food without adding sodium. She's a big fan of the combination of acid and heat. A dash of cayenne with a squeeze of lime is one of her favorite ways of building flavor without salt. She reduces wines, juices, and tomatoes to make sauces, which end up thick and delicious with very little sodium.

The most important thing about reducing your use of salt doesn't have to do with taste, though, says Goldman. It has to do with your brain. She advises people on a low-sodium diet to focus on surprising their palates with new foods they've never had before, or with unexpected textures. By giving the mouth novel experiences, you distract the brain, and Goldman thinks you can retrain your brain to expect a surprise instead of salt.

"It shoots you past the salt problem and into the experience of enjoying your food and trying to figure out what you're eating. The element of surprise gets lost in salt."

There's another thing at work here, too: the concept of tolerance. Just as alcohol drinkers and drug users build up a tolerance for their substance of choice, normal, healthy eaters build up a tolerance for the level of salt they consume. It's very hard to try to back down from this tolerance abruptly, as Goldman had to do. It's much easier to ease yourself off salt one pinch at a time, which is what I should have done with my fresh heirloom tomatoes instead of abandoning salt altogether.

Techniques for Lowering Salt

Campbell's and Frito-Lay have both made efforts to reduce the sodium content of their foods, though Campbell's fired the first shot. Since Campbell's is known for their soups, and soup tastes best when it's salted liberally, they had a lot to lose. Yet they have done a respectable job reducing the sodium in many of their soups as well as launching new flavors with less sodium. In advertising, they tout the use of natural sea salt in their reduced-sodium recipes, not so subtly communicating to us that sea salt can be used to lower sodium.

Frito-Lay has marketed their use of Alberger salt in reducing sodium. This salt has a unique shape and more surface area than regular salt, so it dissolves more readily on the tongue, resulting in a quick, strong hit of saltiness. Using it, Frito-Lay claims, will allow them to reduce the level of sodium across their line of snacks, which includes Lay's, Tostitos, and Doritos.

I have no doubt that the use of natural sea salt did help Campbell's reduce sodium in their soup. And I'm sure Alberger salt played a role in Frito-Lay's sodium reduction success. But I can almost assuredly tell you that neither ingredient on its own was responsible for all of the sodium reduction in the products in which they're used. Whereas you can use about a dozen ingredients to replace sugar, including natural stevia, when it comes to reducing sodium without reducing saltiness, you have to employ more than just one secret ingredient.

When it comes to salt—like fat—there simply is nothing like the real thing.

Less Gravity, Less Sodium

When NASA nutritionists decided that they wanted to reduce the sodium in astronauts' food for its health benefit, they hired Mattson. Our assignment was not just to reduce the sodium, but to cut it in half for every item on the astronauts' menu in order to compensate for typical human behavior in outer space. Most of the condiments the astronauts use to enhance the flavor of their foods are loaded with sodium. They have access to pure liquid salt, but not the crystalline kind. Apparently, if you were to sprinkle salt crystals on food in zero gravity, they would float around and potentially damage some of the precious equipment that accomplishes many things—one of them being getting you home to Earth. The nutritionists figured that if they reduced the sodium of the meals available on board, then a few extra squirts of salt or hot sauce would put the sodium level in the meals right about where they wanted it to be.

Because astronauts travel for extended periods of time, the food that goes with them is sterilized in a process that's similar to canning, so we at Mattson didn't have to rely on salt as a preservative. But we were dealt a blow when the nutritionists told us they were also trying to limit potassium, so we could not use potassium chloride in lieu of sodium chloride. Usually a food developer's first weapon for lowering sodium, potassium chloride tastes salty but has less sodium than salt.

Doug Berg, one of our best and most senior food technologists (also a trained chef), and Samson Hsia, our former Executive Vice President of [Food] Technology, told me they had twenty-nine items to optimize, and each one required a unique approach. The baked beans were fairly easy, for example, "because there's a lot of acid, there's a lot of sweetness from the tomato, molasses. There's savory flavor from the garlic and onion. There's a little mustard, so you get a teeny bit of pungency, a teeny bit of bitter," said Berg. "If you have all five of those tastes, you have a lot more leverage to play around with."

He continued to talk about different entrees, such as the crawfish etouffé and an Indian curry chicken and rice.

"You take the salt out and flavor starts spiking in ways that would not be evident without that salt. You lose the balance. All of a sudden you're tasting acidity, then a sharp herb. Single notes of ingredients. They're kind of competing with each other. When you had the salt in there, you had a nice balance of flavors."

"Salt homogenizes the flavor," said Hsia, meaning this in a good way.

So Berg and Hsia used umami taste enhancers; they pushed acidity higher to compensate for having less sodium; and they enhanced the odor of the salty items by adding appropriate herbs and spices that increase the perception of saltiness. Alberger salt wasn't relevant for the astronauts' meals because everything was premixed, as in a curried chicken and rice dish. This is why Alberger works in Frito-Lay's seasoning mixes, because it sits on the outside of the chips, but not in Campbell's soups because the salt is already dissolved in liquid food.

Reducing sodium is never simple. In fact, it's one of the most challenging things a chef or food technologist has to do. After reading this book, I hope you'll have such a deep understanding of how tastes and aromas work together that you can, if you need to, be as successful as NASA in reducing the sodium in your diet.

No or Low on the Tomato?

Would we all be better off eating the way Jessica Goldman does? The good news is that we don't have to go so far. We can find a happy medium between her diet and the way most Americans eat today. After I had my naked-tomato epiphany and experienced a tomato for what it was supposed to be, I chose not to keep experiencing it that way. I found the temptation of salting my tomato simply too great to resist.

Many people in the food industry believe that taking the salt out of processed foods will simply result in people adding it back at the table, via the shaker. But this notion doesn't hold water (salt pun intended). We don't reach for the salt shaker when we're near to dying from salt depletion, yet we do eat so much more salt than we actually need to sustain life.

Paul Breslin describes what happens when you put salt on a fresh tomato, essentially summing up the superheroism of salt:

For one thing, you've got volatiles that will come off the tomato. You can smell the tomato and its juices, in liquid form. So you may be salting out some of the volatiles. It might smell more strongly after salting it. You'll also make it salty-tasting, which is, of course, desirable. Tomatoes are very rich in free glutamates, MSG, naturally. All of that umami-ness that comes out of a tomato, particularly in the middle mucky part, the gelatinous part of the tomato—the salt will complement the umami-ness and vice versa. And you may be suppressing any bitter notes that are inherent in the tomato so you'll be altering the overall profile. And the degree that bitterness was previously suppressing other flavors in the tomato, like tartness or sweetness, you'll be releasing them from suppression.

He concludes, "That's why salted tomatoes are so good."

Salt

Measured by: Sodium content

Classic Salt Pairing: Salt + Umami
Examples: Chicken soup, bacon
Why it works: Salt enhances the umami and umami enhances the salt

Classic Salt Pairing: Salt + Bitter
Example: Salted grapefruit (very judiciously)
Why it works: The salt suppresses the bitterness inherent in the grapefruit. This allows the sweetness and sourness to come through more cleanly.

Classic Salt Pairing: Salt + Sweet + Sour + Umami
Examples: Barbecue sauce and barbecue seasoning; teriyaki sauce
Why it works: Barbecue sauce is one of the most popular salty-sweet flavors in the world. When it's used on top of meats, chicken, or fish, you get more tastes than just the umami in the meat. American-style barbecue sauce gets sweet from tomatoes, honey, or molasses; salt from salt; sour from tomatoes or vinegar; and umami from tomatoes. Teriyaki sauce gets sweet from fruit juice or sugar, salt from salt, sour from fruit juice, and umami from soy sauce.

Classic Salt Pairing: Salt + Sweet
Examples: Honey-roasted nuts, salted caramel, chocolate peanut butter cups
Why it works: Sweet and salty flavors are popular across a wide variety of foods, but the pure blending of granulated sugar and salt is one of the simplest and most elegant. While there isn't much (if any) honey on honey-roasted peanuts, they are delicious because of the sweet and salty combination.

Aromas Associated with Salt:

Cheese	Celery
Chicken broth	Ocean
Fish	Ham
Seafood	Smoke
Beef	Cured meat

YOU WILL NEED
    2-cup liquid measuring cup
    Boiling water
    4 Lipton, PG Tips, or other black tea bags*
    3 glasses
    Masking tape and markers
    4 tablespoons sugar
    ⅛ teaspoon salt
    3 spoons
    1½ cups cold water
    Saltine crackers for cleansing your palate

DIRECTIONS
1. Pour 13 ounces of boiling water over 4 tea bags in the measuring cup and let the tea brew for 10 minutes. You want to overbrew it so that the bitterness is pronounced.
2. While the tea is brewing, mark the glasses with tape on the bottom. Mark them:
    • Tea
    • Tea + S
    • Tea + S + S
3. Put 2 tablespoons of the sugar in the glass marked Tea + S.
4. Put 2 tablespoons of the sugar and ⅛ teaspoon salt in the glass marked Tea + S + S.
5. Remove the tea bags after the 10 minutes and discard.
6. You should be left with 12 ounces of tea. Equally divide the tea among the glasses so that each glass gets 4 ounces of tea.

---

* Any bitter black tea (such as English breakfast) or green tea will work. I developed the exercise using Lipton tea bags, though, because they are widely available.

7. Put a spoon in each glass and stir until all the sugar and salt are dissolved.
8. Pour another 4 ounces of cold water into each glass and stir.
9. Taste all 3 teas and note how bitter and sweet each one tastes.

DISCUSS

1. You'll notice that the tea (Tea) tastes bitter and the sweetened tea (Tea + S) tastes less bitter.
2. When you taste the tea with sugar and salt (Tea + S + S) you should notice that it is slightly less bitter than Tea + S but it's also slightly more sweet. You've just experienced the superheroism of salt. It thwarts the bad tastes (bitter) and enhances the good tastes (sweet).

Taste: The Bitter-Masking Power of Salt

YOU WILL NEED

½ cup sugar, divided into ¼ cup measures
2 bowls
¼ teaspoon salt
Masking tape and markers
Knife
½ grapefruit for each person tasting
Saltine crackers for cleansing your palate

DIRECTIONS

1. Measure ¼ cup sugar into each bowl.
2. To one bowl, add the salt and mix well. With masking tape, mark this bowl on the bottom so you can tell which one has the salt in it.
3. Cut the grapefruits into wedges.

4. Sprinkle the cut surfaces of half of the grapefruit wedges with sugar.
5. Sprinkle the cut surfaces of the remaining wedges with the sugar and salt combination.
6. Give all tasters one wedge of each of two pieces of grapefruit from *the same piece of fruit* so that the only difference between the two is the salt.
7. Taste the wedge with sugar first. Note the level of sweetness and bitterness.
8. Eat a cracker to cleanse your palate.
9. Taste the wedge with sugar and salt next. Note the level of sweetness and bitterness.

DISCUSS

You'll notice that the wedges of grapefruit with salt on them taste a tiny bit sweeter and less bitter. This is a result of the superheroism of salt.

9

Bitter

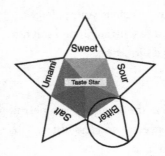

In the years between World War I and World War II, U.S. Army Quarter-master Captain Paul Logan was tasked with the job of stocking military bases, vessels, and soldiers with nutritious food. Everyday dining hall chow was easy. A bit more difficult was developing the field ration, a sturdy kit of food provided to soldiers for eating between bouts of combat. And most challenging of all was developing a form of nourishment for men who ended up in truly desperate conditions, such as those whose planes were shot down or lost in a jungle. Logan wanted these men to have an emergency provision, solely capable of sustaining them for three days and ready to eat without preparation. It needed to be small enough and light enough—no more than four ounces—to fit in a soldier's pocket, since space is at a premium in a plane, boat, life raft, backpack, or uniform. It also had to withstand extremes of temperature, from the frigid cargo bay of a plane and Northern European winters to the warmth of the human body wearing the pocket to the tropical heat and humidity of potential war zones such as the South Pacific islands.

The challenge was also to create a food that tasted bad enough that the men wouldn't eat it in nonemergency situations, but would save it for truly perilous ones. In Logan's words, the food had to taste "a little better than a boiled potato." The solution, he decided, was chocolate, and in 1937, Logan went to The Hershey Company. Sam Hinkle, the chief (chocolate) chemist at the company, accepted the job and created the D-ration, a chocolate bar that Hershey made exclusively for the United States military. Nicknamed the Logan bar, it was never going to win any culinary awards. With a melt point of 120°F, the Logan bar lacked the distinctive hedonic quality of other chocolate, the way it starts to melt the minute you put it in your mouth. In fact, some servicemen with bad teeth could barely eat the Logan bar at all because it was so difficult to bite into. But even hard, waxy chocolate was better than many other wartime provisions. Hinkle's solution for ensuring that the Logan bar wouldn't be traded for cigarettes or girly magazines was a stroke of genius: he developed perhaps the original high-cacao chocolate bar, so bitter it was universally despised—unless you were on a life raft or trapped behind enemy lines with nothing else to eat.

Bitter Medicine

In small amounts, bitter things like caffeine and alcohol can have very pleasant effects. In excessive doses they can be lethal, just like cancer drugs—many of which are bitter, by the way. When cooking with bitter ingredients, you want just enough bitterness to make the dish healthful and complex-tasting, but not enough to kill the dish. Bitter is, essentially, the chemotherapy of taste.

When food was scarce, cavemen had to scrounge for it just to survive, they quickly learned not to eat too much of the stuff that made them sick. They developed conditioned aversions to things like certain bitter foods that upset the stomach or caused diarrhea. It's likely that the foods that made people sick were high in phytonutrients, plant nutrients that have some medicinal benefits. When these bitter compounds are eaten in large quantities, they act like poisons. But it's not pleasant to eat really bitter food in large quantities, so it's a brilliant system: a food that's so poisonous it will make you sick is also so bitter that you won't want to eat a toxic amount of it.

Most people avoid bitter tastes. When we're developing food at Mattson, we are constantly challenged to make (usually healthful) food taste less bitter so more people will buy it. I can't remember a project where our goal was an assertively bitter taste profile, though I'm hoping we get this brief someday. I'm salivating at the challenge.

Biological Basis for Bitter

Sweet and bitter are the two Basic Tastes for which newborn humans have the most robust ingrained responses: sweetness signals that a food contains calories whereas bitter warns that it may contain poison. Most—but not all—poisonous foods are bitter. And most bitter things—but not all—are poisonous in high quantities.

Humans have only one or two taste receptors for sweet, but dozens of taste receptors for bitter because we need to be broadly and instantly aware of stuff that can kill us. Every time you taste something bitter, stop, count your blessings, and be thankful that your poison detection system is functional, as it's the first line of defense for your survival. Tasting a food tells you very quickly whether you should swallow or spit, to avoid ingesting poison, which can kill you.

If you think of taste receptors as a police force, then bitter foods are repeat offenders flaunting their pharmacological power. Bitter swaggers. Bitter preens. It comes across as dangerous and is usually pronounced guilty before being proved innocent.

Bitter Bits

Many things in food taste bitter. Unlike the sour taste, which comes only from acids, bitter compounds include amino acids, peptides, esters, lactone, phenols, polyphenols, flavonoids, terpenes, methylxanthines (caffeine), sulfimides (saccharin), and salts. This is why we need so many different receptors. We need to recognize all bitter substances in order to avoid them at harmful levels.

Coffee is one of the most widely consumed bitter foods in the world, and most people think it's the caffeine that gives coffee its characteristic bitter taste. In fact, only about 10 percent of the bitterness in coffee comes from caffeine. The rest comes from phenolic acids, formed during the roasting process as well as the temperature, time, and method you use to brew the coffee. This makes sense, as decaffeinated coffee can be just as bitter as caffeinated. The bitterness in tea and chocolate (which both contain caffeine) also comes mainly from phenolic compounds other than caffeine.

Your genes influence whether or not a food tastes bitter to you. Remember the chemical PROP, used to test for sensitive taster types? Twenty-five to 30 percent of the population cannot taste it at all. The vast majority of people find it slightly bitter. A quarter to a third of the population finds it unbearable. There are probably receptor genes for other bitter tastes (caffeine, quinine, and so on) that make people experience them differently. Individuals differ widely in our ability to detect bitter tastes, much more so than for the other Basic Tastes.

It's possible that ancient tribes who lived in parts of the world where there was no threat from ingesting a certain type of bitter taste (tea leaves or broccoli, for example) might have lost the specific bitter receptor for the compounds recognized by that receptor. (Animals, including humans, tend to lose or inactivate genes that aren't useful for survival.) One recent study asked adults to identify the predominant tastes they experienced in their diet for a period of a week. Not surprisingly, only 5 to 8 percent of the calories they ate were rated as bitter. This may mean that those of us who live in highly developed nations—where we eat very little in the way of bitter foods—might start to lose more of our bitter receptors. It won't happen in the short run, but it could happen over centuries of cushy living and bitter-free eating.

When Too Bitter Is Just About Right

At Mattson, we always test the prototypes we develop by giving them to consumers. They unabashedly let us know right away if we're kidding ourselves that we've developed a truly Big Idea. We recruit children to taste kids' prototypes and if they make faces like the one on the next page, we go back to the drawing board. For us, this face is the sign of failure. But another company is actually aiming for this reaction to its product, denatonium benzoate, which goes by the trade name Bitrex. In fact, the company uses a picture like this one to market the product. Bitrex is not just really bitter; it's Guinness World Record bitter. Bitrex is the most bitter substance known to man.

Imagine you're developing a household product, such as a liquid soap, that you want to smell like strawberries. If you were smart (because you read this book) you would make it a beautiful bright red color to further communicate that it smells of strawberry, knowing that our eyes are critically important to the perception of aroma. The problem is that bright red, shiny liquids look delicious to young children, who, if they get their hands on this soap and smell strawberry, will put it in their mouths to see if it tastes as good as it smells. If your soap also happens to be poisonous, this poses a real problem. Bitrex to the rescue! Because Bitrex is colorless, odorless, and completely harmless, a tiny amount of it in your strawberry soap will ensure that any child who puts it in his or her mouth will spit it out immediately. And make the bitter face.

Bitter is also added to antifreeze, which has a notoriously sweet taste, in the form of denatonium benzoate (Bitrex), now a critical ingredient in the formula to make it unappealingly bitter to kids and animals. If you're trying to quit biting your nails, you can buy products containing denatonium benzoate to paint onto the tips of your fingers. A few naughty nibbles at your cuticles and you'll be dashing to the bathroom to rinse your mouth out.

Bit Parts

For my birthday one year, Roger offered to cook me anything I wanted for dinner. Normally, I do most of the cooking in our household, mostly because I like to, but also because Roger doesn't. He grills meat much better than I, and he can sear a mean lamb chop, but although he's a passionate eater, he is just not very interested in cooking.

So I thought I'd make it easy on him for my birthday. I requested that he procure (not cook) crabs, one of the comfort foods from my childhood. Cooking live crabs is a task best left to professional cooks, those born and raised in Maryland, and the nonsqueamish. Roger is none of the above. Because we live in California, flying my beloved blue crabs across the country would have required a lot of advance planning and politically incorrect air travel for the doomed crustaceans. So I decided to rough it and settle for our local Dungeness, with a side of oven-roasted potatoes and a simple salad to round out the meal.

While Roger was shopping, someone gave him the excellent advice to make the salad with bitter greens to balance the sweetness of the crabmeat. As a non–green food eater, Roger took this quite literally. He picked out one fresh head each of frisee, endive, and chicory. And just for good measure, he added a fourth lettuce: a Treviso radicchio so bitter I went to the store the next day to find out what it was (so I could avoid it in the future). Roger tossed the salad with tree-green extra virgin olive oil, freshly squeezed Meyer lemon juice, salt, and pepper. We sat down to eat and I took a big forkful of the beautiful produce on my plate. That was my first and last taste of the salad.

"Don't you like it?" Roger asked, more than a little crestfallen. "Oh, it's perfect," I said. "Bitter. Very bitter. Just what you were going for. Maybe a little too bitter?" I offered as I flushed my mouth out with water and a big hunk of bread before returning to the crab. It was simply too much of a good thing.

That's the rub with bitter foods. Just like chemotherapy, they play a very important role in small doses, requiring restraint so as not to overwhelm. What Roger might have done was build a salad from sweet greens like butter lettuce or mâche, and used a small amount of the bitter greens as supporting actors. We can tolerate few bitter foods as the star of the show.

Daily Dose of Bitterness

As much as people love coffee, most who drink it doctor it up. Seventy-five to 80 percent of the population perceives coffee as bitter, from just a bit to off the charts. Yet the 25 to 30 percent of the population who are Tolerant Tasters may experience black coffee as lacking bitterness entirely. You know these people. They're the ones drinking espresso as if it tastes like water.

Adding Taste Star counterpoints like milk and sugar, to coffee, balances the bitterness with other Basic Tastes. In fact, the daily act of adding milk and sugar to coffee is one of the best examples of how we intuitively balance tastes without thinking. If you use a dairy product like 1%, 2%, whole milk, or cream, in addition to lowering the bitterness of the coffee, you're also adding fat to it. Fat adds a desirable, creamy mouthfeel that coats the tongue and makes the coffee taste less bitter. The way this works is that the bitter compounds in the coffee are diluted into the fat phase of the milk. This makes them less able to reach bitter taste receptors. When you consider using fat as a counterpoint to bitter, it's interesting to note that fat is also a contender for Basic Taste status, as we'll see later in the book.

One of the goals of the Taste Star is to help you think about *all* tastes when you're cooking, not just the ones that come easily to mind (salt, sugar, sour). When you taste a dish, envision the star, and think about what's missing by considering each point in the star. Does it need salt? Should it be sweeter? Should it have a touch of acid? Would umami make it richer? Does it want bitterness? Most people try to avoid bitter. But the talented chef knows how to use bitterness to add complexity to a dish. By *complexity* I mean adding another counterpoint, as well as making the resulting food more challenging than it normally is. Our lazy palates easily accept sweet foods, but sweetness with a "just about right" level of bitterness makes you stop and think, *Hmmmmmm, there's something interesting going on there.* Bitter plays this complexifying role.

Bitter also makes extremely salty or sweet foods less so. In other words, it suppresses salt and sweet.

Bitter Balancing Act

I used this knowledge one night while cooking up a batch of barbecue sauce on my stovetop. I had smoked a beef brisket (unsuccessfully, due to my lack of patience), which took a few more hours than I'd planned for. I was frantically trying to get everything finished and ready to serve before our dinner guests arrived. I stopped following the measurements on the barbecue sauce recipe and started improvising. When I finally got the sauce reduced to a viscosity that I liked, I dipped my spoon in to taste. Too sweet! With very little time to get this sauce right, I considered the Taste Star, which led me to ingredients on the four other points. I added a few dashes of soy sauce to give a bit more umami, some seasoned rice vinegar to pump up the sourness without adding much aroma (because the volatile aroma of lemon, for example, doesn't belong in an American barbecue sauce). I tasted again. Still too sweet. And now, thanks to the soy sauce and seasoned vinegar (which both contain sodium), too salty! The only taste that was going to get me out of this mess was bitter. I opened my pantry. Bitter. Bitter. I needed something bitter that would marry well with my tomato, honey, and vinegar sauce. I grabbed the container of unsweetened cocoa powder and voilà! The bitter cocoa powder balanced my otherwise sweet, sour, salty, and umami sauce. Perfect Taste Star harmony.

Bitter Means Good for You in Moderation

We're learning more about bitter every day, but we already know that this Basic Taste usually indicates some kind of pharmacological function. Aspirin is bitter and has well-known pharmacological benefits. Ibuprofen is bitter and functions as an anti-inflammatory drug. Tea is bitter and high in antioxidants.

Many vegetables such as greens, and fruits such as pomegranates and cranberries, taste bitter because of polyphenols, flavonoids, isoflavones, terpenes, and glucosinolates, lumped together under the term *phytonutrients*. And guess what? These are the compounds responsible for giving produce the ability to help lower the risk of cancer and heart disease. Imagine if we could enhance the phytonutrient content of our foods to make them even more healthful. Just as The Tomato Project's Harry Klee wants to create the most delicious yet hearty tomato, we could use traditional breeding techniques to create plants that have more of the healthful compounds. Unfortunately, it's difficult to do this without increasing the bitterness, which people don't want to taste.

The word *bitter* could use a public relations agency to improve its image.

Public Relations Campaign to Improve Bitter's Image, Take 1:
Everyone's atwitter over the complex taste of bitter.

Bitter Taste Confusion

Cranberries have a unique type of mouthfeel, called *astringency* or *tannin*, which gives the sensation of having your tongue dried out. In fact, cranberries are so tannic, astringent, and bitter that they are barely tolerable in naked form. Dried cranberries (or Craisins, the brilliant trademark that Ocean Spray coined) are infused with sugar to balance the bitterness that's naturally occurring in the fruit. The compounds that make cranberries bitter may be responsible for their function in helping avoid and alleviate urinary tract infections.

Many tannic or astringent foods are bitter, but tannin and astringency are not experienced through bitter taste receptors. Instead, they act on the trigeminal nerve that carries touch information to the brain. Although we generally refer to our perception of tannin and astringency as tasting bitter, physiologically speaking it's a feeling, not a taste. To experience the tannic astringency of grape skins as bitter, do the Feeling Tannic? exercise at the end of the chapter.

Many people also confuse the Basic Tastes bitter and sour, probably because many sour foods are also bitter. Grapefruit and cranberry are two great examples. Both have an assertively sour taste with some strong bitter notes, especially in the pith of the grapefruit and the skin of the cranberry. Sour tastes sharp and pungent. Think of vinegar or lime juice. Bitter tastes unpleasant. Think of unsweetened espresso, tea, or chocolate. Because these are both Basic Tastes, it's really hard to describe them without using the words sour or bitter. To experience the two, I suggest you do the sensory exercise Differentiating Bitter from Sour at the end of the chapter.

A Bitter Choice

I make the world's best Brussels sprouts. I blanch them lightly, cut them in half, and then sear them, flat side down, in bacon fat on the highest setting my gas-burning range can achieve. I toss them with a spritz of seasoned rice vinegar, crisp cubes of bacon, a spritz of fish sauce, and a suspicion of sea salt. Each bite is a balanced combination of bitter, sweet, sour, salt, and umami, but Roger refuses to eat them. In his sensory world, Brussels sprouts are simply too bitter, and no amount of culinary makeup hides that.

People who taste PROP (the bitter marker compound) as bitter are likely to perceive vegetables such as Brussels sprouts and kale as bitter, too, and thus are less likely to eat them.

If you live with a bitter rejecter, I hope this chapter endows you with taste empathy. This person is not trying to make your life as the family cook more difficult, but is probably just a HyperTaster with genetic intolerances for specific bitter tastes. But the fact that some members of your family won't eat bitter doesn't mean you shouldn't try to get them to eat more vegetables. Vegetables are clearly good for us and there are some easy ways to tame bitterness, some of which you instinctively already know.

Taste Empathy

A vicarious experience of another person's taste type. Taste empathy is necessary to truly appreciate and understand how wildly different we all experience the sense of taste.

The easiest way to balance the taste of bitter foods is to add counterpoint tastes, such as sweet, sour, and salt. To Brussels sprouts, add salt and, yes, even sugar. Don't be afraid to add anything sweet, such as honey, maple syrup, agave, or juice to a dish that's out of balance for your dearest bitter rejecters. You can also enjoy the same bitterness-balancing effects of sweetness without the calories. Aspartame (Nutrasweet, Equal) and sucralose (Splenda) both do the job effectively. If adding sugar to vegetables sounds like cheating, you'll quickly get over it when you learn that a study showing that "cheating" by adding as little as 5 percent sugar not only resulted in higher liking scores for the sweetened cauliflower and broccoli, but these same test subjects liked *unsweetened* cauliflower and broccoli better from that point forward. Think of sugar as training wheels for the appreciation of bitter vegetables, not as cheating.

Another way to increase the sweetness of a food without adding sugar is to slowly caramelize it in a bit of oil in the oven or on the stovetop. Even sulfury, bitter cauliflower can become crave-ably sweet in the oven.

Simply cooking vegetables reduces their bitterness. Some of the volatile aromas will flash off when you steam, boil, roast, or otherwise cook a vegetable such as broccoli. The sulfurous smell that may stink up your kitchen means that you've liberated the aroma from the cells of the vegetable into the air. While unpleasant smells, like sulfur, don't contribute a bitter taste, they do exacerbate it, so minimizing icky smells through cooking results in a sweeter, less bitter overall perception.

I like to add salt and sugar in the form of seasoned rice vinegar (rice vinegar with sugar plus salt added), which also adds sourness. With a few dashes, you can easily add three Basic Taste counterpoints. Remember that salt suppresses bitterness, but it also releases other more desirable flavors from suppression. So even if all you add is a topical shake of salt, you'll be helping tame the bitterness in more than one way.

The chef's golden rule is taste, taste, taste. If you taste your broccoli raw, and then taste it again after you blanch it, you'll get a sense of how blanching affects the bitterness and aroma. If you stir-fry it in a wok, taste it after a minute. The bitterness will be likely to have changed again. Add soy sauce (salt and umami) and sugar (sweet), then taste it. Finally, adjust it with acid— maybe a squeeze of lemon or a dash of vinegar—and taste it again. The more you do this, the better you'll start to understand what works best to balance the flavors of your food.

• •

Public Relations Campaign to Improve Bitter's Image, Take 3:
Bitter. Complex. Sophisticated.

• •

Making Bitter Less So

Mary Tagliaferri, president and chief medical officer of a start-up biotech company called Bionovo, knew that the bitter taste of her company's herbal hot flash remedy might sabotage its chance at success. Dr. Tagliaferri has a background in traditional Chinese medicine in addition to a degree from the prestigious medical school at the University of California, San Francisco. Her company's goal is to bring together the best of both Eastern and Western medicine. In 2010 and 2011, Bionovo was putting its herbal remedy through the U.S. Food and Drug Administration's process for drug approval. If the patient trials were successful, Bionovo could market Menerba, its proprietary blend of herbs, as a pharmaceutical, one that would require a prescription.

Phase 1 of Menerba's FDA drug trials proved that a low dose was safe. For Menerba's Phase 2 testing the dosage was increased, but the resulting formulation was too voluminous to deliver in a pill or capsule, so they decided to create a powdered beverage mix like Theraflu. After months of working unsuccessfully with a supplier to develop a drink that would mask the bitter taste of Menerba, Bionovo called us at Mattson.

Our assignment was to develop a dry beverage mix that the patient would add to water and drink twice a day. When Dr. Tagliaferri and her colleagues brought their samples to us, I didn't know what to expect. Most of the functional foods we develop aim to deliver delicious taste first, with health benefit a distant second. Menerba was an entirely new thing for us. Women who were desperate enough to seek out medical treatment for hot flashes were going to be less concerned with the taste of their medicine than someone buying a beverage at the local convenience store.

Nonetheless, I was unprepared for the taste of Menerba. In fifteen years of developing new food and beverage products, I had never tasted anything quite so alarming. Menerba hit the palate, immediately, with an acrid, burnt aroma—a result of how the herbs were processed and refined. This burnt note then led into a whole-mouth bitter taste. The finish, the taste that remains in your mouth after you swallow, held the lingering bitterness of the functional ingredients and an almost moldy, earthy flavor. It was wholly unpleasant.

We strategized on what flavor to develop. Dr. Tagliaferri wanted a lemon or orange flavor because both are widely appealing, but we knew that citrus would be extremely difficult because consumers aren't expecting bitterness from lemon or orange drinks. We first considered a spicy chai tea, since tea is inherently bitter and the spices would mask some of the unpleasant aromas, but chai isn't exactly a mainstream flavor, so we abandoned it after one round of formulation. We then settled on a fruit flavor that would be acceptable with a bit of background bitterness: cranberry. It was the perfect choice because consumers would expect a cranberry beverage to have a bit of an edge. We had also noticed during the development of the chai tea idea that versions of the drink with vanilla were rounder, less bitter. So we tried a very low, undetectable level of vanilla in the drink. It was astonishing: the vanilla helped mask the bitterness of Menerba. In addition, we added just a touch of bitter-suppressing salt—at a threshold level that no one could detect as salty—and the beverage came into focus. Finally, we added a spoonful of sugar to help the medicine go down.

I have since used vanilla as a secret weapon to make just about anything taste better. I'm convinced that the vanilla latte is the most popular coffee-house drink because vanilla masks the bitter flavor of coffee in a way that hazelnut, caramel, and chocolate just don't do as well. It's also why chocolate cake recipes call for vanilla, and why many chocolatiers add a low level of vanilla to their confections. You need to use only a tiny amount of vanilla to take advantage of its bitter-masking effects. Interestingly, vanilla is the only flavor that doesn't seem to have an upper limit for tolerability. At high levels, where most other flavors have already gone from good to better to way too intense, vanilla doesn't stop savoring delicious.

Bitter Genes

If you scrape the inside surface of your cheek for a DNA sample, Monell scientist Danielle Reed could test it and tell you whether you will find the chemical PTC bitter or not. Unfortunately this test doesn't determine whether you find other things like tea, spinach, or cocoa bitter, or what type of food choices you'll make. But we may soon understand more about how our genes determine our reaction to bitter tastes. Reed can foresee a future when many people eat foods that are developed to appeal to their specific genes. Instead of marketing their chocolate as 41%, 62%, or 82% cacao, companies like Scharffen Berger will simply indicate which genotypes their Nibby Dark Chocolate bar will appeal to, based on the degree of bitterness each can handle.

I prefer to envision a future where consumers will relish the thought of eating something that assaults their bitter taste receptors. My future doesn't require genetic testing or avoiding certain foods. It simply requires a change of perspective in which we consider bitter to be an indispensable tuning knob for balancing flavor. When you scrunch up your face at bitterness, it's likely that the bitterness is out of balance. So if you're eating a plate of Brussels sprouts and you just can't get past the taste, add counterpoints to bring it into balance. Sprinkle on a bit of sugar. Add a dash of salt. Spritz on a bit of lemon or vinegar. Or better yet, toss the sprouts with other vegetables such as sweet potatoes, carrots, or caramelized onions. The next time you make them, use fewer carrots and more sprouts. Eventually you'll find yourself craving a bowlful of them—alone—specifically for the energizing, stimulating taste challenge that bitter provides.

Bitter

Measured by: Phenolic content, alkaloid content, etc.

Classic Bitter Pairing: Bitter + Sour
Example: Tea with lemon
Why it works: The acidity of the lemon softens the tannic bitterness of tea. Add sugar or honey and you've added another counterpoint, further balancing the flavor.

Classic Bitter Pairing: Bitter + Fat
Example: Coffee with cream
Why it works: Coffee is both bitter and sour. Adding fat rounds the sharp edges of both. The bitterness of coffee is best masked by a combination of fat and sugar.

Classic Bitter Pairing: Bitter + Salt
Example: Grilled, steamed, or stir-fried green vegetables such as asparagus, broccoli, Brussels sprouts, or kale
Why it works: Salt suppresses the bitterness of green vegetables. Add a bit of fat, too, for a side dish that won't scare HyperTasters away from the table.

Aromas Associated with Bitter:

Coffee	Herbal
Chocolate	Brewed
Green vegetables	Hoppy/hops
Sulfur	Red wine
Smoke	Alcohol
Acrid	Metallic

Taste: Adjusting the Bitterness of Coffee

Of course you've tasted coffee before. But this time stop, slow down, and taste coffee again as if for the first time. Pay attention to all five Basic Tastes. Use all five of your senses. Each time you taste, evaluate:

1. Appearance
2. Aroma
3. Taste
4. Texture
5. Sound

YOU WILL NEED

A cup of black coffee

Sugar

Creamer or milk (your choice of fat level)

DIRECTIONS

1. Take a sip of coffee. Notice how bitter it is. Consider the acidity of the coffee. How sour is it?
2. Add a small amount of sugar to the coffee. How much? Just about the right amount. Now taste it. Notice how the bitterness has changed.
3. Add a little bit more sugar to the coffee. Taste it again. Consider how the bitterness changed this time.
4. Now add the creamer to the coffee, about 1 tablespoon at a time. Stop and taste it after each additional tablespoon. Notice what the creamer did to the bitterness of the coffee.
5. Also notice what the creamer does to the sourness of the coffee. The pH of coffee is lower than milk (coffee is *more* sour than milk). When you add milk to coffee you raise the pH level of the resulting drink, making it less sour.

Taste: Feeling Tannic?

YOU WILL NEED

A bunch of seedless grapes in the darkest color you can find.
Red will work, but black is even better.

DIRECTIONS

1. Wash the grapes. Put one in your mouth and suck on it. Do not bite it.
2. *Without using your teeth*, smash the grape up against the roof of your mouth. Try to remove the interior flesh from the skin of the grape.
3. *Without using your teeth*, suck on the grape parts until the flesh is gone and you are left with just the grape skin.
4. Begin to chew the grape skin, slowly. Keep chewing until you start to experience a certain change in mouthfeel.

OBSERVE

The sensation that you are getting is the drying of astringency. It comes from the tannin in the skin of the grape.

Taste: Differentiating Bitter from Sour

YOU WILL NEED

Masking tape and markers
2 ramekins or cups
1 lemon for each person tasting
A citrus juicer
1 clear glass liquid measuring cup
Table salt

Spoons
Sharp knife or vegetable peeler
Saltine crackers and cold water for cleansing your palate

DIRECTIONS

1. Put a label that says Juice + Salt on one ramekin.
2. Juice the lemons into the measuring cup. Divide the juice evenly into the 2 ramekins.
3. To the Juice + Salt ramekin, add 1 pinch of salt for each lemon you juiced. Stir well to dissolve.
4. Cut off bits of lemon rind so that each taster gets enough to chew for a few seconds. Be sure to include the white pith as well as the yellow skin. Set aside.

TASTE

5. Taste the plain juice first. What you are experiencing is mainly sour. There might be a tiny bit of sweetness or bitterness in the juice, but the big, bold, intense flavor you are tasting is sour.
6. If there was any bitterness in your lemon juice, the touch of salt should mask it. Taste the Juice + Salt. Now we can say for sure that what you are tasting is almost purely sour.
7. Now put the lemon rind in your mouth and chew on it for 30 seconds before spitting it out. The taste you're experiencing now is bitter. It is strong and intense like sour, but it is of a different quality.
8. Cleanse your palate with crackers and water and go back and forth a few times to really understand the difference.

10

Sweet

When I was a child in the 1970s, women drank Tab, the beverage of choice for fashion-conscious, forward-thinking, good-looking people. At least that's what advertising suggested. Tab was the first sugar-free soda to be marketed by the Coca-Cola Company and was sweetened with saccharin, a substance that's five hundred times sweeter than sugar. Saccharin is so intensely sweet that Coke had to use only a minuscule amount of it to craft a beverage with the same sweetness as a regular Coca-Cola. The company marketed it, cleverly, as having only one calorie.

But not every cola-drinking woman switched from regular Coke to Tab. Some women who wanted to switch to a lower-calorie cola just couldn't bring themselves to drink Tab because of the taste.

The two characteristic Basic Tastes in cola are sweet and sour. You probably don't think of cola as tasting sour, because much of its acidity comes from the carbonation, or carbonic acid, which feels tingly as well as tasting sour, and the prickly bubble irritaste distracts you from thinking about the sourness on your tongue. On the other hand, most people do think of cola as tasting sweet. Once you mess with the sweetness of a drink like Coca-Cola, you're left with a different thing altogether. Sucrose, otherwise known as sugar—which was used to sweeten cola in the seventies—is the prototypical sweet taste. It is the purest, cleanest expression of the Basic Taste sweet. Sugar-free Tab had its work cut out for it.

Today we have a few more sugar-free alternatives to Tab. There's Diet Coke, sweetened with aspartame (200 times sweeter than sugar), Diet Coke with Splenda (600 times sweeter than sugar), and Coke Zero, which is sweetened with a blend of aspartame (200 times sweeter than sugar) and acesulfame potassium (abbreviated aceK, 180 times sweeter than sugar). There was even a short-lived Coke product called C2 that was sweetened with a blend of high fructose corn syrup, aspartame, acesulfame potassium, and sucralose. Although I have no insider knowledge, I'll bet right now they're working on a few more sugar-free cola versions with new and different sweeteners.

Why do beverage companies keep pursuing sugar-free soft drinks? There are two reasons: The first is that no other sweetener tastes exactly like sugar. It's not that saccharin was a bad choice for Tab. It's that trying to replace sugar with *anything* else is problematic. The second is that our genetic and anatomical differences mean we all perceive sweetness a little bit differently. Each sweetener has its own *sweetness flavor display* or *sweetness profile*, terms we use to describe how we perceive its unique sweetness. This makes it nearly impossible to create a sugar-free drink that everyone will accept.

Sweetness Profiles

If you were to taste the myriad sweeteners available today, such as sugar, sucralose, aspartame, acesulfame potassium, stevia, and honey in succession, you'd understand how complex the seemingly simple Basic Taste sweet is. When you do the Sweetness Profile exercise at the end of the chapter, pay attention to the three phases of what you're tasting. The first phase is what we call the *up-front* taste, which occurs at the very beginning. This refers to how fast the sweet taste registers in your brain. You'll notice that the speed with which you taste each sample's sweetness varies greatly. The second phase is what happens after the taste registers in yout brain. This is the *middle* of the taste profile. How much does the sample fill your mouth? Does it taste thin? Bitter? Where does the taste hit on your tongue? The last phase, or the *finish*, happens after you swallow and the taste starts to fade away. Many artificial sweeteners have a very long finish, which means that their taste remains in your mouth for a long time.

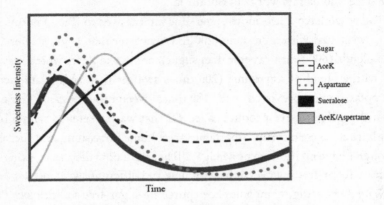

A Simulation of Sweetness Time and Intensity Curves

Look at the graph of the sweetness time and intensity curves. Notice that sugar (the thickest line) takes a while to come on. It builds up in your mouth much more slowly than aspartame, which knocks you over with an immediate sweetness. Notice also that sucralose is still screamingly sweet long after the sweetness of sugar, aspartame, and aceK have peaked and started to decline. This difference in the timing and intensity of how you experience sugar and *nonnutritive sweeteners* is why many people find that the nonnutritive sweeteners don't taste quite right.

One of the ways that companies like Coke get around the different timing and intensity curves is by blending more than one sweetener together. Doing this allows one sweetener's curve to fill in the peaks and valleys of another's, smoothing them out. The blend of aceK and aspartame that's used in Coke Zero moves the sweetness hump a bit more in the direction and shape of the sugar curve. This is probably why Coke Zero was one of Coca-Cola's most successful new product launches. Its sweetness profile more closely resembles "the real thing."

A Sweet Touch

Even if you were able to prove that a blend of the sweeteners aceK and aspartame makes diet cola A taste exactly as sweet as sugar-sweetened cola B, you'd still come up short, because sugar adds more than just the Basic Taste sweet. It also contributes to mouthfeel.

Sugar sweetness is measured using an instrument called a refractometer, which measures the degrees of Brix, or the amount of solids in a food, which correlates with sweetness. The higher the °Brix, the higher the solids, the sweeter the taste. Think of how Concord grape juice (for example, Welch's) feels in your mouth: sweet, sour, and a tiny bit tannic. Now consider grape jam; it's a completely different textural experience although it has the same flavor profile: sweet, sour, and a tiny bit tannic. The difference between grape juice and grape jam is the amount of sugar and water, and the stiffness of the gel formed by pectin. The juice has much less sugar, and hence a lower °Brix. Welch's Concord grape juice has a Brix of 15° while Welch's grape jam has a Brix of 65°. The fourfold increase in Brix is one of the things that makes the jam feel much thicker on your tongue.

Brix refers to both sweetness and thickness, which are related. You've experienced increasing Brix if you've ever oversweetened a cup of hot tea or coffee. As you add sugar one teaspoon at a time, it slowly dissolves into the tea. The more sugar you add, the thicker the tea gets. If you got lost in thought and continued to add teaspoon after teaspoon, eventually you'd end up with a thick, syrupy drink. With sugar, an increase in viscosity accompanies an increase in sweetness.

When you take the sugar out of a cola and replace it with an infinitesimal amount of a nonnutritive sweetener such as aspartame or saccharin, you lose all the Brix from the sugar and hence you lose some of the mouthfeel. Artificial sweeteners are used at such a low level that they don't increase the amount of solids in a drink, which is also why they're called *nonnutritive*. They're used at incredibly low levels and they don't provide any nutrition or calories. The result is a beverage with less body. Tab, Diet Coke, and Coke Zero all lack the satisfying mouthfeel that bulkier sugar provides.

Sweet Signals

My cat G.G. is as sweet as sugar. She'll sit in my lap and cuddle with me for hours during a movie. If you offer her your hand, she'll lick it clean with her sandpaper tongue, whether it needs a washing or not. She's generally a canned food kind of cat who turns up her nose and leaves the room if I dare offer anything but her beloved Fancy Feast. But one morning I heard G.G. crunching on something hard that she was thoroughly enjoying. When I asked my sweetheart kitty, "What are you eating, G.G.?" she stopped chewing and spit out the head of a mouse.

A few hours later, still disturbed by the image, I wondered: *Does she really find dead mice delicious?* They're furry, full of bones, and raw. Yuck. I can understand that a mouse might taste good if I were starving and built a fire to make a rosemary rotisserie mouse. But a raw, stringy mouse? Sounds horrible.

If G.G. were able to talk, she might have the same reaction watching me eat Swedish Fish, a chewy confection that's neither Swedish nor marine. These fish-shaped, fruit-flavored candies are almost pure sugar, and I adore them. They return me to my childhood, when my mother used to let us buy them at Morrow's Nut House before we went to the mall movie theater. But G.G. is a carnivore that has evolved to eat only meat. Somewhere inside that peanut-sized kitty brain of hers is a reward system that says *"YUM! Raw mouse! Delicious! Sustaining! Protein! Fat!"* Sugar holds no appeal for her. Since cats don't need it for nutrition, they have lost the ability to taste sweetness.

Humans, on the other hand, crave sugar.

Sensory Snack

Hippocrates would diagnose diabetes by tasting the patient's urine. If it tasted sweet, he had his positive indicator. Today we have much more sophisticated ways to test for sugar in the urine, no doubt to the great relief of medical lab technicians.

Sweet Response

When you put something sweet in your mouth, the insulin level in your body spikes. This is what is called a cephalic phase response to what your body assumes (from the sweet taste) will be incoming carbohydrates. It's essentially a reflex. Your tongue signals *sweet* to the brain, which communicates to your gut: *Food is coming; prepare for work*. Then acidic digestive juices start flowing.

What happens when the brain senses something sweet that doesn't provide calories? This question has led to much research into the effects of nonnutritive sweeteners such as aspartame and sucralose. Some people believe that sugar-free sweeteners interrupt the natural taste/eat/reward system. Others think these sweeteners can be used specifically for their deceptive sweetness, to allow you to make do with fewer calories without giving up sweetness in your diet. Dozens of studies failed to provide consensus on whether use of these sweeteners leads to weight gain or loss. Nonnutritive sweeteners probably make consumers prefer higher sweetener levels in foods and beverages, but we don't know whether this leads them to eat less and lose weight.

I believe that you can teach yourself to like things less sweet. Once your palate has become used to a norm, you have to slowly reset it. For example, I drink a cup of hot tea every morning when I wake up. I used to sweeten it with two packets of high-intensity sweetener. Then one day while doing research for this book, I realized that I was drinking the sweetness equivalent of four teaspoons of sugar each morning and decided to cut back. When I first tried to go lighter on sweetness, I felt I was missing something. Each morning I dreaded making my one-packet tea, which no longer gave me the same sensory stimulation I was used to. So I decided to try a different tack. I switched from sweetener to honey, which adds not just the sweet taste, but also a lovely floral aroma and body from the Brix that honey provides. I was tricking myself by replacing some of the sweet taste with other senses: smell and touch. Today I'm drinking a much different cup of tea: less sweet but equally satisfying.

Retraining your palate to like things less sweet is one of the secrets to eating a more healthful diet. It will be startling at first, but once you figure out how to swap one type of sensory stimulation (sweet taste) for another (smell texture) and appreciate it for its own sensation, you won't miss anything at all. The key is developing awareness and appreciation of sensory stimulation beyond sweet. If you find yourself drinking or eating something really sweet, you can kick your lazy palate into gear by using other points on the Sensory Star, as well as other Basic Tastes.

Balancing Sweet

Regardless of where you live, how you were brought up, or whether you're a HyperTaster, Taster, or Tolerant Taster, in general, you like the taste of things that are sweet. You may not like intensely sweet foods—like Swedish Fish candy—but it would be odd for anyone to taste sugar and say it is unpalatable. We're highly tuned to three sweet molecules—sucrose, fructose, and glucose—because they provide quick, easy sources of energy (otherwise known as calories). Back in the days when our ancestors were trying to feed their families by hunting and gathering, they needed all the calories they could get. Hunting for game was hard. Refining starches was time-consuming. But fruit was a no-brainer. It was sweet and you didn't have to strip, skin, cook, mill, or ferment it. If you found a piece of fruit, you'd found a good source of food that immediately fed your hunger.

People so universally love the sweet taste that making foods sweeter is an easy, but lazy, way to make them more palatable. Many Americans would rather eat overly sweet foods than explore the other tastes that come out when sweetness levels are lowered. But this is beginning to change. Milk chocolate in the United States used to be available in only one level of sweetness: Hershey's. Nowadays you can find dark milk chocolates or semisweet milk chocolates in supermarkets, and premium chocolatiers such as Michael Recchiuti and Vosges are growing in popularity. I have a huge fondness for Hershey bars from my youth (spent within driving distance of their chocolate factory), but my chocolate tastes have grown up. So have Hershey's. In 2005, they purchased artisan chocolate maker Scharffen Berger, which offers an "extra rich milk chocolate" that tempers its sweetness with the bold bitter counterpoint of 41% cacao.

I think of sweetness as one point of the Taste Star that requires a counterpoint, or several, to be really crave-able. The most logical counterpoint to sweetness is sourness, and these two occur together in perfect harmony in nature. When fruit ripens to the perfect balance of sweet and sour, it's usually at its nutritional peak.

Chefs, food producers, and winemakers diligently mind the balance between sweet and sour. We refer to this as the *Brix-to-acid balance*. Back to the example of Welch's grape juice and jam. The 15° Brix juice has a sour pH of 3.35: just the right level of tartness to make it refreshing. When Brix gets higher, as it is in the 65° Brix jam, you need more acidity to cut through the sticky sweetness. Welch's grape jam has a pH of 2.9: much more sour than the juice (remember that lower pH equals higher sourness). Without the extra acidity, the jam would taste flat and cloying. Balance is more important than the individual pH or Brix numbers.

I love sweet white wines such as riesling, kerner, gewürtztraminer, and sylvaner, which need a specific Brix-to-acid balance. In my opinion, people who turn their noses up at sweet wine have just not tasted sweet wine with a proper Brix-to-acid balance. When the acidity is high enough to balance the sweetness in a wine, you don't perceive it as too sweet. Just delicious. The same holds true for red wines, which must have a Brix level that balances their acidity. If there's little bitterness, as well, these wines taste flat and flabby and can be uninteresting to drink. Getting the Brix-to-acid balance *just about right* is critical to making something that people want to keep eating or drinking.

Sweet Alchemy

Cinnamon has a familiar aroma that you might describe as spicy, warm, musky, or woody. You might also describe cinnamon as sweet, although anyone who has ever gotten a mouthful of plain cinnamon knows it's not really sweet. This confusion between the senses is a result of associating cinnamon with breakfast pastries, cookies, cakes, and other sweet foods where cinnamon is used. When you smell cinnamon, even in the absence of sugar, or taste it plain without anything else, you remember sweetness, which most foods with cinnamon have. The same is true of vanilla, nutmeg, and cocoa. In fact, cocoa is assertively bitter, yet due to our experience with it in sweet foods, we sometimes say cocoa smells sweet. Because of this association between certain aromas and the Basic Taste sweet, you can make something taste sweeter by upping the amount of vanilla or cinnamon in it. You're not adding sweetness, just the perception of it.

When sugar is heated to a high enough temperature it starts to melt, then undergoes a startling transformation. The transformation of granular white crystals to golden brown syrup is one of the most satisfying types of alchemy a chef can perform. Caramel is browned sugar. To caramelize something else, then, is to brown the naturally occurring sugar in that food. For example, you can bring out the natural sugars in an onion by slowly cooking it until the sugars brown. Other wonderful things happen when you cook sugars. Lovely nutty, roasty aromas develop. Maillard browning is created when sugars are heated in the presence of amino acids and browning itself creates hundreds of flavorful compounds, many of which are some of the most desirable flavors known to man—browned meat, baked bread, and roasted coffee.

Sensory Snack

You've probably heard of real estate agents baking a batch of cookies during an open house, trying to make the home they're selling smell, well, *homier.* It may not be such bad advice after all. Researchers have proved that consumers on a tight budget will spend more after being exposed to the aroma of chocolate chip cookies.

Sweet Relief

The taste of sugar helps relieve pain. Doctors all over the world employ the sweet taste to mitigate pain from many medical procedures, from taking blood to circumcising baby boys, who will cry less and show less distress if they have a sweet taste in their mouth during the surgery. This is such a well-known phenomenon that Western parents used to give babies sugar water to soothe them and some parents in less developed countries still do. The downside to using sweetness as a pain reliever for kids is that babies who are given sugar water early in life like sweetness more when they are older and sometimes grow up to eat more sweets, get more cavities, and become overweight.

On average, kids like sweeter things than adults do. When you mature, you lose your love for exorbitantly sweet candies, gums, and drinks, although some people still love the supersweet stuff they did as kids. It's likely that these sweets lovers aren't drinkers, because the amount of alcohol you drink and how much sugar you eat are inversely correlated. There is a belief that the sensory satisfaction from alcoholic beverages scratches the same itch that sugar does. The more you drink, the less likely you are to eat candies, cookies, and cakes. But if you quit drinking, you'll start to crave these sweet things. Clearly, drinking alcohol and eating sweets are both pleasurable activities, but whether they work the same way in the brain is unclear. It's no coincidence that sugar consumption skyrocketed in the United States following the prohibition of alcohol in the 1920s. The US Federal Trade Commission's *Report on Sugar Supply and Prices* stated "prohibition is one of the causes of the greatly increased number of candy stores, ice cream establishments, and soft-drink manufacturers." The temperance movement aimed to reduce crime and domestic violence, but it ended up causing an unintended consequence: encouraging America's consumption of the Basic Taste sweet.

Sweet

Measured as: Degrees of Brix, or °Brix

Classic Sweet Pairing: Sweet + Sour
Examples: Hard candies such as Life Savers and Jolly Ranchers
Why it works: In order to make sugar harden into candy, you can't
have a lot of other things in the candy. One thing that does work
is acid, which luckily also fits the fruity flavor profile of most hard
candies. A hard candy without the proper level of acidity is cloying
and unpleasant. And a hard candy with too much acid can be
unpleasant if you're over a certain sour-loving age.

Classic Sweet Pairing: Sweet + Fat
Examples: Pastries, cake, cookies
Why it works: When you're craving something soft and sweet,
you really want to treat yourself. What better way to prolong this
sensory experience than to reach for something that's made with
a lot of tongue-coating fat? The fat carries the sweetness on your
tongue and makes the experience last longer.

Classic Sweet Pairing: Sweet + Bitter
Example: Chocolate
Why it works: Chocolate without sugar is like Laurel without Hardy,
Bogey without Bacall, or tomatoes without salt. It's just not the same.

Aromas Associated with Sweet:

Vanilla	Fruit
Cinnamon	Nutmeg
Berries	Caramel
Honey	Bread
Maple syrup	Cake
Apple	Butter
Chocolate	

Taste: Sweetness Profile

YOU WILL NEED

    1 quart room-temperature water (8 ounces water in each cup or
        bowl), plus more for cleansing your palate

    4 cups or bowls

    1 packet Equal

    1 packet Splenda

    1 packet Truvia

    2 tablespoons sugar

    Saltine crackers and water for cleansing your palate

    Paper and pens or pencils

DIRECTIONS

1. Pour the water equally into the cups. Dissolve the sweeteners into separate cups of water, one at a time. It will take a bit longer to dissolve the sugar. Make sure it's all dissolved before starting the exercise. All of the other sweeteners will go into solution much more easily.

2. Taste these sweetened waters, considering the following things:
   - Notice how quickly you detect the sweetness.
   - Count how long it takes for the sweetness to fill your entire mouth.
   - Notice how full the sweetness is in your mouth.
   - Count how long the sweetness lasts.
   - Write down any other flavors you taste.

3. Taste the sugar water. After the sweetness is gone, eat a cracker and drink some plain water to cleanse your palate. Write down your reactions.

4. Repeat for the Equal water.

5. Repeat for the Truvia water.

6. Repeat for the Splenda water.

OBSERVE

- Remember that the only Basic Taste you are experiencing in each of these samples is sweet.
- Why do they *taste* so different?
- It's because each one of them has a different sweetness curve. Review the Sweetness Profile graph. Does this make more sense to you now?

Taste: Rating Brix and Acid

YOU WILL NEED

    Coca-Cola or Pepsi
    Orange juice
    Gatorade, any flavor
    Frappuccino bottled coffee beverage
    Milk, 2%
    Brewed coffee, black
    Cups for everyone who is tasting
    Pens and paper for tasters to write down their answers
    Saltine crackers for cleansing your palate

DIRECTIONS

1. There are two tasks in this exercise.
2. The first task is to sample each beverage, then rank the beverages in order of increasing sweetness.
3. The second task is to sample each beverage again, then rank the beverages in order of increasing sourness.
4. The answers appear at the end of the "Sour" chapter, page 243.

OBSERVE

1. Keep in mind that sugar and acid balance each other. Some beverages that are very sweet may also be very sour, but the tartness is masked by the sweetness.
2. Discuss your answers with the other tasters.

11

Sour

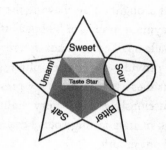

Curt Mueller played basketball at the University of Wisconsin in the early 1960s, after which he headed not to the National Basketball Association but into the lab. As a budding pharmacist, he decided to tackle some of the physiological issues he had faced on the court. During particularly strenuous games his mouth would dry out, resulting in a dreaded condition called cottonmouth. Drinking water was problematic. Basketball moves so fast that water sloshed around uncomfortably in his stomach. Mueller tinkered with sprays that effectively solved the problem, but when he tested them, athletes had trouble managing the clunky bottles.

In the seventies Mueller launched a chewing gum "that stimulates saliva," its package promised, for all types of athletes, solving the dry mouth problem in a clever way. Mueller's gum was so incredibly sour it would make your mouth fill with saliva. He called his concoction Quench.

One of the ways we describe delicious food is by saying that it's mouth-watering. Sour is the Basic Taste that makes our mouth water most. This happens when a supersour food enters the mouth—saliva to the rescue! Saliva rushes in to try to manage this huge change in acidity in the mouth. The more sour the food, the more saliva floods the mouth. Once there's enough saliva to dilute the sourness, the waterworks stop. That's why Quench was so brilliant: as you kept chewing it, it kept releasing more acid, which kept the saliva flowing.

Mueller marketed his product as a "sport gum." Quench did nothing to alleviate thirst, but that was okay, because it wasn't supposed to rehydrate the body with fluids. It was just supposed to eliminate the dry feeling in the mouth. With a mouthful of spit brought on by chewing the squintingly tart gum, athletes could continue playing, running, or biking a little longer without a break.

Human saliva is made up of compounds that represent all five of the Basic Tastes: glucose, which provides sweet; urea, bitter; sodium chloride, salt; and glutamate, umami. But these Basic Tastes are present at levels that are too low for us to detect consciously. Saliva is very slightly acidic, but when we're healthy, we don't think of it as tasting sour, because we are accustomed, or adapted, to the taste of our mouth.

Sour (Patch) Kids

My grandparents used to have an apple tree in their backyard in suburban Baltimore. At a young age I'd climb it to pick apples fresh from the branches, sometimes too early in the season. That didn't stop me from eating the budding fruit, though. I loved the bracing sourness of the flesh, crisp and only slightly sweet.

If too many days went by between visits to Grandma's, I might miss the window of time when the apples were perfectly ripe. It was an elusive sliver of autumn between too-tart green apples and overripe red ones that fell off the tree and rotted on the ground. This narrow stretch was when the apples tasted best to the adults, when the fruit had the most perfect balance of puckery acidity and sweet juice. At the precise moment when they are neither too tart nor too sweet, apples also offer their peak nutritional value.

Some scientists believe that the biological reason we're born with a sourness detection system is so that we'll end up eating more nutritious, riper fruits and vegetables. Apples that are too tart don't taste good, at least not to babies and adults. The sourness stops us from eating the fruit until it's ripe—and sweet—enough that we derive the maximum nutrition from it.

Humans can detect sourness from birth. Babies are acutely sensitive to it. If you need proof (and have an evil streak), you could give an infant a wedge of lemon and watch him scrunch up his face and push it away. Having a built-in sourness rejection response makes sense, because babies' developing systems cannot tolerate acidity as well as those of adults. Too much acid in children's diets can harm their developing teeth and upset their developing digestive systems.

What doesn't make sense is that kids from about five to nine years of age prefer sour foods at acidity levels that both babies and adults reject. To see this for yourself, just visit a candy store. The candies aimed at children in this age range are not chocolates, caramels, or nuts, but sour candies, some of them promising super, atomic, crybaby, or "toxic" levels of tartness.

Charles Darwin noticed this bizarre phenomenon in 1877 while chronicling the development of his own kids. He wrote:

> It looked to me as if the sense of taste, at least with my own children when they were still very young, was different from the adult sense of taste; this shows itself by the fact that they did not refuse rhubarb with some sugar and milk which is for us an abominable disgusting mixture and by the fact that they strongly preferred the most sour and tart fruits, as for instance unripe gooseberries and Holz apples.

Modern-day researchers at Monell conducted tests with mothers and children to find out what levels of sourness each preferred. They scientifically proved what the makers of Warheads, Sour Patch Kids, and Tear Jerkers have known all along. Using gelatin desserts that they spiked with higher and higher levels of acidity, they found that "a striking difference emerged between children and their mothers in their preferences for the extreme sour gelatin." This difference is what allowed me to enjoy unripe apples and Charles Darwin's children to prefer "abominable" sour fruits.

What could explain the fact that newborn babies and adults reject sour foods that kids between the ages of five and nine crave? There are many things we don't know about the sour Basic Taste and the biological reason for it.

In addition to spoilage, sour is the primary taste of fermentation, which is often used to preserve food. In fact, fermentation is a form of spoilage. After the active cultures develop in sauerkraut and yogurt, for example, the result tastes sourer than before. During a trip to San Francisco in my early twenties, I remarked to my dining companion that the bread tasted tart and made a face as if it had spoiled. "It's sourdough, you idiot," my friend reminded me. I simply wasn't used to my bread tasting sour. Now that I live in San Francisco, if my bread isn't slightly tangy, I miss the acidity.

The Science of Sour

We use the words *sour* and *acid* in our daily vernacular. Sour tastes come from acids.

In the food industry, we measure acid level—or acidity—on a scale from zero to fourteen, with seven as a midpoint. The numbers indicate the concentration of hydrogen ions in the food or beverage, abbreviated as pH. Foods that have a pH below 7 are acidic. Anything above 7 is alkaline. Because 7 is the midpoint, a food that has a pH of 7 is called *neutral*, neither acidic nor alkaline. Water has a neutral pH of 7.0. That's one of the reasons it's perfect for cleansing the palate. Water helps return the mouth to a more neutral pH. When you're evaluating food (and anything, for that matter), neutrality is always a good place to start.

The hardest thing to remember about pH is that a lower pH means a higher level of sourness. A higher pH means a lower level of sourness. For example, a lime has a pH of about 2.0. Milk has a pH between 6.0 and 7.0. The lower pH 2.0 lime is more acidic than the pH 7.0 milk. I know, I know. This makes having a discussion about pH confusing. I wish I could tell you that we have no problem communicating pH level in the food development world, but this "higher means less" system is counter to what we know about numbers. When we're talking casually in the lab, we often screw it up and have to correct ourselves.

The pH of Common Foods, Bodily Fluids, and Household Items

The pH scale is exponential. Each number is 10 times more or less acidic than the next one.

Not edible	Drain cleaner	14	
	Bleach	13	
	Soapy water	12	Alkaline
	Ammonia	11	
	Milk of magnesia	10	
	Baking soda	9	
	Blood, some milk	8	
	Water	7	Neutral
	Saliva, urine, milk	6	
	Black coffee	5	
	Tomato juice	4	Acidic
	Orange juice	3	
	Lemon juice	2	
Not edible	Gastric acid	1	
	Battery acid	0	

Just to add more confusion to the pH scale, it's not a normal linear scale. It's exponential. That means that a pH of 4.0 is ten times more acidic than a pH of 5.0. And a pH of 4.0 is ten times less acidic than a pH of 3.0.

Many studies have tried and failed to draw some correlation between an acid, a food, its pH, and how sour it tastes. Using a variety of different acids, you could make up sour water drinks that all measured the same pH—for example, pH 3.0—but when you taste them, they will have different levels of sourness. In other words, if you taste pH 3.0 lemon water (which contains citric acid) versus pH 3.0 vinegar water (which contains acetic acid) versus pH 3.0 lactic acid water (the type of acid in cheese) with your nose pinched shut, you will notice that each has a different level and quality of sourness.

. . . .

Using 3 different acids dissolved in water, if you . . .
>   matched their pH, they would taste different.
>   matched their taste, their pH would be different.

. . . .

Conversely, if you were to tinker with your acidic waters until they all tasted exactly as sour as one another, their pH measurements would be likely to fall within a fairly wide range of about a whole pH point. Believe it or not, at the same pH level, vinegar water tastes sourer than hydrochloric acid water. Of course, this would be a very low concentration of acid, because hydrochloric acid at a higher concentration can corrode metal and burn human skin.

At the other end of the pH scale lives alkalinity. Alkalinity is actually an indication of a substance's ability to buffer an acid or decrease acidity (which equates to an increase in pH). Here's an everyday food example that will help explain how this works.

When you brew a cup of thick, rich black coffee, the two Basic Tastes will be an assertive bitterness and a slight acidity. You can balance the sourness with milk.

That's because black coffee falls within the acidic range of the pH scale, with a measurement around 5.0. Milk, half-and-half, and cream have a pH that's higher, usually between 6.0 and 8.0. If you add milk to more acidic coffee, your resulting beverage will have a pH higher than 5.0. In other words, you move your creamy coffee up the pH scale—and this means it tastes less acidic. Milk makes coffee less acidic. Of course, other things are going on. If you add milk that contains fat, you're also changing the mouthfeel, which helps to tame some of the compounds that make coffee bitter.

The change in coffee pH from a slightly acidic 5.0 to a less acidic, say, 6.0 might seem small, but moving coffee up the pH scale by one point means that it's ten times less acidic. So adding cream to coffee results in a creamy beverage with ten times less sourness.

A lot of the foods humans love are acidic and that makes heartburn a chronic illness for some. The physical problem is that your throat tissues weren't built for the pH 1.0 gastric acid that bubbles up into your chest. To remedy this, you can take antacid tablets, which help buffer the acid to a higher pH, making it less likely to burn and cause discomfort.

Feeling Sour

At low levels, acids taste sour to us. At high levels, they also have a tactile component: an irritaste. The trigeminal nerve, which carries sensations of pain and touch, can detect odorless acids when you sniff them. If you've ever gotten a deep noseful of vinegar, you may have experienced this yourself. I like to reduce balsamic vinegar into a sweet-and-sour syrup to drizzle on strawberries, ice cream, and grilled proteins such as salmon or pork. But if I heat it up too quickly and the volatiles start to waft off, I can feel it in my nasal passages. Even if I couldn't smell it, I could detect it.

A Sour Situation

In the Mattson food lab, we have a complicated relationship with the sour Basic Taste.

We rely on a high level of acidity to keep microbes away from and out of foods. For example, the acidic kick of canned salsa and tomato and marinara sauces allows food companies to market them without refrigeration. Many shelf-stable foods use acidity for preservation. The reason they don't taste puckeringly sour is because they're balanced with other tastes and aromas.

FDA's Classification of Foods Up to pH 8.0

	Food	pH (average)	How We Use pH or Modify pH
Low-acid foods, according to FDA food-processing regulations	Milk	7.4	Lower acidity in coffee
	Human saliva	6.5–7.5	Floods the mouth to lower acidity
	White rice	6.35	Foil for acidic sauces, add vinegar to make sushi rice
	Potatoes	6.1	Increase acidity by adding sour cream
	Cheddar cheese	5.9	Serve with acidic fruits, pickles, pair with tomatoes.
	Bread	5.55	Slather with mayo, mustard, jam, jelly to increase acidity
	Cucumbers	5.4	Increase acidity by pickling
	Bananas	4.85	Eat raw or use to lower acid in strawberry smoothies
		4.6	
High-acid foods, according to FDA food-processing regulations	Tomatoes (fresh)	4.5	Add acidity and umami to salads, pasta, pizza
	Mayonnaise	4.35	Add tang (and fat!) to sandwiches, salads
	Tomatoes (canned)	4.1	Add acidity and umami to pasta
	Red Delicious apples	3.9	Perfectly balanced from nature: eat raw
	Sorrel	3.7	Adds a fresh note to salads and soups
	Dill pickles	3.35	Counterpoint to low-acid foods like sandwiches, cheese, soup
	Jams/jellies	3.3	Add fruity freshness by spreading on baked goods
	Rhubarb	3.25	Add sugar and bake into pie
	Vinegar	2.7	Add tang to salad dressings, sauces; use for pickling
	Cranberry juice	2.4	Needs to be sweetened to be enjoyed as juice or otherwise
	Lemons	2.3	Add acidity to fish, pasta, dressings, sauces, etc.
	Limes	1.9	Add acidity to salsa, alcoholic drinks, etc.

Source: http://www.fda.gov/Food/FoodSafety/FoodborneIllness/FoodborneIllnessFoodbornePathogensNaturalToxins/BadBugBook/ucm122561.htm

Winemakers and jam and preserve makers balance acid with Brix (sweetness). Condiments, which are almost always high in acid, are also balanced with salt (think mustard and hot sauce) and sometimes sweet (think ketchup and Thai sweet chili sauce). Juices are high in acid but balanced by sweetness. Bartenders, known as *mixologists* these days, are also excellent practitioners of balancing sour with other Basic Tastes. They use the bitterness of alcohol as a third counterpoint to sweet and sour. Take the most popular cocktail in America: the margarita. Without the bitterness of the tequila, the drink would be cloyingly sweet and disconcertingly sour.

Sour and Salt

Some acids taste salty. And some salts taste sour. Often we confuse them as scientists have proved. We tend to misperceive salt as sour and sour as salt because of the way they're detected at the taste receptor level.

When I went to the Center for Smell and Taste in Florida, I was tested for how well the taste buds on certain areas of my tongue worked. Researchers "painted" a cotton swab dipped into a sour solution across different parts of my tongue. In some places, it tasted sour. In others, it tasted salty. This is because the sour and salt Basic Tastes are both perceived the same way: through a channel. Sweet, bitter, and umami are perceived in a hand-and-glove fashion: the molecule fits into the receptor.

Perhaps because of their similar detection mechanism, sourness can enhance saltiness and vice versa. Delfina's Craig Stoll uses acid to drive the intensity of his food.

"You've got salt and acid that you keep amping up in a vinaigrette until it hits this crescendo of flavor. Where we want it," he says, "if you had to sum up our food, it would be depth and brightness," by which he means acidity. "Most of our stuff is balanced with some acidity somewhere. And some of it's hidden. We finish a lot of our sauces with vinaigrette. We do a leg of lamb where we'll make a lamb jus and reduce it and on the pickup, we'll warm it up and we'll add a quarter or half ounce of sherry wine vinaigrette." Stoll does the same thing with roast chicken and silken olive oil mashed potatoes. But he cautions that if you can taste the vinegar, his chef has added too much. He's not using it for its flavor, but for its taste—the Basic Taste sour, which sparks food alive. Acidity is an easy way to literally make food more mouthwatering. Within a small, sour window, that is.

Sour

Measured by: pH

Classic Sour Pairing: Sweet + Sour
Example: Chinese sweet-and-sour chicken, pork
Why it works: If this dish is done well, it's irresistible. If it's not, the sweetness can become cloying. The sweetness plays off the subtlety of the meats, and the sourness gives the whole dish—often fried, then sauced—a freshness that lightens it up.

Classic Sour Pairing: Sour + Fat
Example: Salad dressings/vinegar and oil
Why it works: The harsh acidity of vinegar needs a taming counterpoint. While the other Basic Tastes work great, sometimes fat is needed to further round the edges. Salad dressing is one of those cases.

Classic Sour Pairing: Sour + Salt
Example: Pickled vegetables
Why it works: Pickles are preserved with acidity. Adding salt to sour pickles not only balances the tartness but also suppresses any bitter flavors that might be present in the vegetable you're pickling, such as radishes or beets.

Classic Sour Pairing: Sour + Hot + Salt
Examples: Tabasco, pickled jalapeños, sriracha
Why it works: Hot sauces are preserved with acidity. Because they're used in such small amounts, you want them to deliver a big punch of flavor. Acid and salt help give the chile pepper this added punch. They also balance out one-dimensional heat that might otherwise be too strong on its own.

Aromas Associated with Sour:

Orange	Vinegar
Lemon	Fermented
Lime	Pickled
Grapefruit	Yogurt
Yuzu	Cultured
Tamarind	Rhubarb

Taste: Rank the pH of the Beverages

YOU WILL NEED
    Coca-Cola or Pepsi
    Orange juice
    Gatorade
    Frappuccino bottled coffee beverage
    Milk
    Brewed coffee (no cream or sugar added)
    Saltine crackers and water to cleanse your palate

DIRECTIONS
    The objective of this exercise is to place the beverages in order
    of increasing sourness, or decreasing pH. (They are the same
    thing!)

The key is on page 243.

Taste: How Brix Changes the Perception of Sour

YOU WILL NEED
    Masking tape and marker
    4 glasses
    2 cups of water, plus more for cleansing
    2 cups lemon juice (freshly squeezed is best, but any kind will work)
    1 nonreactive saucepan
    Liquid measuring cup
    6 tablespoons sugar
    Spoon
    Saltine crackers for cleansing your palate

DIRECTIONS

1. Using the tape and marker, mark the glasses 1 through 4.
2. Combine the water and lemon juice in the saucepan and stir. Measure ½ cup into glass number 1 and place it in the refrigerator.
3. Heat the remaining lemon-water mixture over medium heat for about 2 minutes—do not boil. You want it just hot enough to dissolve sugar, not to bubble or boil (which would concentrate the mixture).
4. Add 2 tablespoons sugar and stir until dissolved. Pour ½ cup of the mixture into glass number 2 and place it in the refrigerator.
5. Stir another 2 tablespoons sugar into the liquid until completely dissolved. Pour ½ cup into glass number 3 and place it in the refrigerator.
6. Stir the remaining 2 tablespoons sugar into the liquid until completely dissolved. Pour into glass number 4 and place in the refrigerator.
7. Let lemonades chill for about 1 hour before tasting.

TASTE

- Taste the unsweetened lemonade (number 1) first. Notice how sour it tastes.
- Cleanse your palate with saltines and water.
- Taste the next three lemonades in order of increasing sweetness, cleansing your palate between tastings with saltines and water.

OBSERVE

- Notice how the sugar totally changes the way you perceive the sourness.
- All 4 glasses have the same level of sourness: in other words, the same pH value. The only difference is the level of sugar, or °Brix.

Taste: Answers to Rating Brix and Acid

Part 1 Answer:

In order of increasing sweetness, the beverages stack up as follows:

Beverage	°Brix
Brewed coffee, black	2.2°
Gatorade, any flavor	6.5°
Coca-Cola or Pepsi	10.8°
Orange juice	12.6°
Milk, 2%	13.2°
Frappuccino bottled coffee beverage	17.6°

Part 2 Answer:

In order of increasing sourness, the beverages stack up as follows:

Beverage	pH
Milk, 2%	6.54
Frappuccino bottled coffee beverage	6.51
Brewed coffee, black	5.08
Orange juice	3.90
Gatorade, any flavor	2.93
Coca-Cola	2.67

Taste: Isolating Acid and Alkaline Tastes

This exercise will give you an understanding of what different pH levels translate to in the mouth. You will be making three different saltwater solutions. The reason we use salt water is that the alkaline ingredient, baking soda, contains sodium. We have to add the same amount of sodium to the two other samples to equalize the salty taste across all three samples.

YOU WILL NEED

Masking tape and marker

3 glasses

Measuring spoons

Table salt

Liquid measuring cup

Hot water

Baking soda (which is pure sodium bicarbonate)

Distilled white vinegar

Saltine crackers and water for cleansing your palate

DIRECTIONS

1. Mark one glass "Neutral pH," the second glass "Alkaline," and the third glass "Acidic."
2. Make the neutral salt water: Add ⅛ teaspoon salt to the "Neutral" glass. Add 1 cup hot water and stir until dissolved.
3. Make the alkaline salt water: Add ½ teaspoon baking soda to the "Alkaline" glass. Add 1 cup hot water and stir until dissolved.
4. Make the acidic salt water: Add ⅛ teaspoon salt to the "Acidic" glass. Add ¼ teaspoon vinegar. Add 1 cup hot water and stir until dissolved.
5. Taste the neutral pH water first.
6. Cleanse your palate with saltines and plain water.
7. Taste the alkaline water next.
8. Cleanse your palate with saltines and plain water.
9. Taste the acidic water next.
10. Cleanse your palate with saltines and plain water.

OBSERVE

1. All three samples will taste salty.
2. The acidic water will taste sour.
3. The alkaline water will taste slightly soapy, slightly odd. That's because we don't eat many foods in the alkaline range.

12

Umami

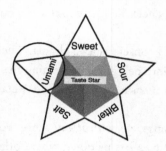

I believe that there is at least one other additional taste which is quite distinct from the four tastes. It is the peculiar taste . . . arising from fish, meat, and so forth.

Kikunae Ikeda, 1909

Chef Stan Frankenthaler spends many of his waking and semiwaking hours thinking about food.

"I get in trouble all the time with my wife for daydreaming about food," he says.

He's been cooking since he was a child, when his parents would allow him to concoct elaborate meals for the family. In his teens and throughout his time at the University of Georgia he worked as a professional cook, and he went to work in restaurants after he graduated. In the early 1980s he went to cooking school at The Culinary Institute of America, finishing first in his class.

Afterward, Frankenthaler headed to Boston and jumped behind the stoves of some of the city's most famous hotels and restaurateurs, including Le Meridien, Lydia Shire, and Jasper White. His cooking earned him James Beard Award nominations for Best Chef Northeast. By the time he opened his own places, The Blue Room and Salamander, he had become known for fusing American cuisine with Asian flavors and wrote a cookbook of Asian-inspired recipes.

Umami is the fifth Basic Taste and it is at the core of Asian cuisine. Yet Frankenthaler, an Asian kitchen veteran, had not heard of this word until the early 1990s. Like most chefs, he knew *that taste*, but he had not had a word for it.

"It was a coming to consciousness, almost. There was a definition—oral definition—that came into being, for something that I'd always done. My experience was, 'Oh, now it has a name,'" he said. Finally, he had a word for *that taste*.

English is a rich language with many more words than we use in daily life. Yet English sometimes lets us down where other languages crystallize a concept with a single word. The French word *terroir* (pronounced tare-WAH) is a perfect example. Terroir, which literally means "soil" in French, communicates so much that it's hard to believe that seven letters can contain it all.

Terroir refers to the specific conditions in which wine grapes (most usually, but also other agricultural products) grow. Geography influences the climate in which the grapes go from sprout to ripe fruit. For example, warm weather can result in sweeter grapes, which can mean bolder, higher-alcohol wines. Geography also accounts for differences in soil, culture, and farming techniques, which inform winemaking methods. Basically, terroir communicates the essence of what makes a grape taste the way it does, and the reason that the grape yields a wine that tastes the way it does.

The French *get* wine. They have the perfect word that enables their understanding of it, whereas we English speakers need a paragraph or so to explain the same concept.

Many Japanese understand umami in the same way the French understand terroir. They *get* it. English speakers haven't grown up with the word, so we don't use it in conversation as we do the four other tastes. Nor has the word *umami* made its way into colloquial food phrases such as "my *sweet*heart," "a *bitter* pill to swallow," "*salt* of the earth," or "*sour* grapes." Umami is a difficult concept to describe and an even harder taste to identify.

The best English words we have to express umami are savory, meaty, brothy, and full. But these words don't do it justice. Describing umami without using the word *umami* would be like trying to describe the taste of salt without using the word *salt*. You could say salt tastes briny, metallic, minerally, stony, or savory, but in reality the Basic Taste salt is none of these things. These words describe flavors that usually occur in tandem with salt. Salt tastes salty and there's just no other way to describe it accurately. This is also true of the four other Basic Tastes. Bitter tastes bitter. Sweet tastes sweet. Sour tastes sour. And umami tastes . . . umami. The reason we say it tastes meaty or brothy is because umami is the primary taste in meat and broth.

The umami taste is thought to signal protein, since many foods high in umami are proteins. And in fact, malnourished, protein-deprived human research subjects prefer higher levels of umami than those who are well fed. Some scientists still don't think umami should be considered a Basic Taste. I do, though. It's something we can clearly detect on our tongue. It's distinct from the four other Basic Tastes. We have a receptor for it. And it adds craveability to foods in a way that other things can't. Most important, in the food lab, when we think about umami in relation to the four other Basic Tastes, it helps us balance flavor. That's enough for me to give it a point on the Taste Star.

Oomami Oomph

Umami makes things taste more delicious, fuller, and rounder. It's the deep, rich sensory difference between raw meat and cooked meat. Ground beef tastes (hopefully!) fresh and clean. It's fairly bland unless you doctor it up with condiments that bring it to life, as with steak tartare. But a cooked steak—on its own with nothing more than a char—becomes a succulent treat, oozing with flavor that's salty, fatty, beefy, and something else. That something else is umami.

Umami is what you're looking for when you dip steamed rice and briny raw fish into soy sauce at a sushi bar. Not too much, unless you want to be branded An American for your excess. The fermented soy adds a depth of flavor that complements the mild freshness of the fish. If you didn't mind the horrified looks, you could salt a piece of sushi or sashimi and you'd enhance the flavor of it, but you'd be missing the oomph of umami that soy sauce delivers beyond saltiness.

Umami is what you add when you sprinkle grated Parmesan cheese on your pizza or pasta. Aged Parmesan cheese is packed so full of umami that it almost deserves its own category. It adds a richness to Italian food that makes us crave it.

It's important to note the distinction between umami and salt. Many of the umami-rich ingredients I mentioned above (meat, soy sauce, Parmesan cheese) are also salty. This isn't necessarily a coincidence. Scientists believe that the sodium part of compounds such as monosodium glutamate, disodium inosinate, and disodium guanylate (other umami-rich salts) interacts with the glutamate to make things umamier. Regardless of how this happens, it's important that you differentiate the taste of umami from that of salt. The easiest way to do this is to complete the Isolating Umami exercise in this chapter, which uses the prototypical form of umami: monosodium glutamate (MSG). When I tell people outside the food industry that we're going to taste MSG plain, they tend to freak out. They worry about getting headaches, a symptom of Chinese restaurant syndrome, an ailment that supposedly follows ingestion of MSG although there is no research that validates this. The way to think about MSG is the same way you think about salt. They're both out there, you consume them every day, and it would be odd if you ever ate them plain, unless you were doing an exercise from a book on the science of taste.

Also like salt, MSG and the other forms of umami occur naturally in foods. Glutamates are found in all sources of protein including meat, cheese, and poultry, plus certain vegetables such as tomatoes, mushrooms, and soybeans. While salt and umami are two very distinct taste qualities, they enhance each other. Both salt and umami occur naturally in foods and both are refined and used as a seasoning. MSG is used mostly in Asia but also sometimes in the West, mostly without consumers' or diners' knowledge.

In 1909, Kikunae Ikeda, a professor in Japan, identified monosodium glutamate as a distinct taste that differed from the four other Basic Tastes and founded a company called Ajinomoto to manufacture and sell MSG as a seasoning. There's not much difference between buying refined, crystallized salt to shake onto your food and buying MSG to use as a seasoning, but for some reason we demonize MSG. I believe this has something to do with our lack of understanding of umami. At reasonable levels MSG—again, like salt—is not to be feared. As writer Jeffrey Steingarten notably quipped, "If MSG is so bad for you, why doesn't everyone in Asia have a headache?"

Most cultures eat foods containing glutamate, some more than others. One study measured the average "daily dose" of glutamate consumed by residents of different countries. The average US intake was 550 milligrams of glutamate (per kilogram of food) per day. Contrast this to Asian countries such as Japan and Korea, where the intake was 1,200 to 1,700 milligrams per day—two to three times that of the United States.

Americans' ability to detect umami has been proved scientifically to be just as good as that of Japanese subjects, although this research tested detection level, not identification: in other words, people could taste it whether or not they could come up with a word for it or a way to describe it. This testing was done in the 1980s, well before the concept of umami was even starting to make its way to the mainstream. We could detect umami long before we knew what it was.

It wasn't until 2000 that a scientist identified the receptor on our tongue that responds to it. The fact that this discovery happened within the borders of the United States certainly helped its cause. Americans could finally claim a bit of umami history as their own. The West became increasingly conscious of umami with the rise in popularity of Asian cuisines. But most cultures have been seasoning their food with umami-rich ingredients for thousands of years. Archaeological excavations of ancient Roman sites have even found evidence of *garum*, a fermented fish sauce used as a seasoning in the time of Julius Caesar just as Asian cultures use fish sauce today.

The Fifth Taste on All Five Continents

The arsenal of umami implements includes Parmesan cheese, Worcestershire sauce, bacon, Vegemite, mushrooms, tomato ketchup, anchovies (the source of umami in a Caesar salad), and Maggi sauce.

David Chang is the chef-owner of Momofuku Ssäm and other restaurants in New York City. Years ago he noticed that Italians and Spaniards lauded their domestic hams while Americans ignored their own equally artisanal southern American cured pork legs. Chang's reverent treatment of Virginia hams at Momofuku Ssäm is worthy of a patriotic culinary medal. Beyond ham (which is loaded with glutamate), his food is decidedly influenced by Asia. More recently he saw another parallel, this time between soy sauce, the reigning king of umami in Asia, and Maggi sauce, which is used in Europe. When he opened his French-like restaurant Má Pêche, he put a bottle of Maggi sauce on each table, a move so clever and lacking pretension that I liked the restaurant even before my food arrived. Born in Switzerland "to bring added taste to meals," Maggi today is owned by Nestlé and sold around the world. It has the same thin viscosity as soy sauce and delivers the same deep, rich flavor of umami. But while soy sauce gets its oomph from the brewing process, Maggi unabashedly lists disodium inosinate (also known as IMP) and disodium guanylate (also known as GMP), two other MSG-like compounds added solely for their umami taste.

We are exposed to the taste of umami well before we're conscious of it, or of anything, really. Human breast milk is high in umami, far higher than cow's milk. But our first experience with umami is in utero—amniotic fluid contains glutamates. We're literally floating in an umami broth until we enter the world.

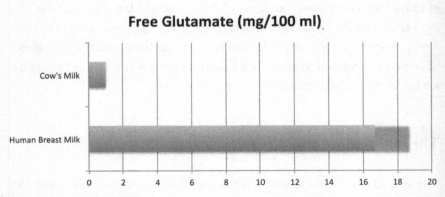

Source: International Glutamate Information Service

How Umami Works

When the large protein molecules in foods are broken down into smaller molecules, they become more flavorful and develop umami. This breakdown is usually a result of cooking, fermenting, drying, or aging. Imagine sucking on a huge cherry tomato that's two inches in diameter. You'll get very little of the taste of it while it's whole. You don't really know how it savors until you chew it down into smaller pieces, when the intense tomatoey flavor is released. This is a good analogy for when meat is raw or cheeses are young: their protein molecules are big and stingy with their flavor. When meat is aged and cooked, or when cheese is aged, the protein breaks down into smaller pieces. The smaller the pieces, the more flavor they offer up.

When I first discovered that there was a huge selection of aged cheeses on the market beyond Parmesan, I went a bit nuts and I bought a huge half-wheel of three-year-old aged Gouda from a local cheese shop for a dinner party. This cheese was a deep, dark, almost orange color, with crystalline crunchies in it. Over the three years it had been aging, the big protein molecules had been breaking down into smaller, more flavorful ones. These crystals were amino acid clusters indicating age and, as a result, huge umami taste. It was an intense, concentrated cheese that tasted so savory it was almost meaty. The lesson I had not yet learned about food that's really high in umami is that a little bit goes a long way. Instead of eating huge hunks as you do with younger cheeses such as mozzarella, my dinner party guests ate tiny shards of it, commenting on the amazingly deep flavor. A year later I still had a huge piece of the Gouda on hand. Each time I entertained, I served richer and richer pieces of it as it continued to age, until I eventually burned myself—and my immediate circle of friends—out on my then-four-year-old aged Gouda.

The vast majority of the cheese that's sold in the United States is semisoft, young varieties such as mozzarella and cheddar. Cheese develops umami as it ages, so these young cheeses are like children: a mere glimmer of what they could become in their maturity. As a result, we eat pizzas and tacos and sandwiches with cheese that provides very little taste beyond salt and of course a fatty mouthfeel. If we were to swap some of that low-flavor, nubile cheese for an umami-rich veteran, we could get away with using a lot less of it. This would be a win-win solution for the diner: you'd ingest fewer calories that deliver more sensory input.

Source of Umami	Umami Comes From . . .	Other Flavors, Tastes, and Sensory Characteristics That Accompany It
Soy sauce	Fermentation, aging	Salty, wine-like characteristics
Tomato paste	Ripening and cooking	Sweet, salty
Worcestershire sauce	Fermentation	Sour, fruity, brewed, mildly salty, small amount of heat
Vegemite	Fermentation	Pronounced saltiness, earthy, beer-like, deep chocolate color
Fish sauce	Fermentation	Salty, fishy, funky
Kelp	Naturally occurring glutamates	Seafood-y, salty, vegetal
Ketchup	Ripening and cooking	Sweet, salty, warm spices
Anchovies	Fermentation	Salty, seafood-y, fishy
Dried tomatoes	Drying	Sweet, sour
Dried mushrooms	Drying	Earthy, vegetal
Bacon and ham	Curing	Meaty, salty, porky
Cheese, especially Parmesan	Ripening, aging	Salty, nutty, meaty, milky, sweet
Fresh tomatoes	Ripening	Sweet, sour

The Beer Goggles of Taste

An interesting and confounding thing about umami is that humans love—crave—foods that are high in umami, yet we don't really enjoy the taste of it on its own. When you do the Isolating Umami exercise you'll notice that the taste of pure umami in water isn't exactly yummy. In fact, it's a bit odd.

Even though it's an odd sensory experience on its own, when umami occurs naturally in a food (or is added unnaturally), it functions as the beer goggles of taste. When you ingest umami-rich molecules, everything you eat with them becomes more beautiful: meaty, salty, flavorful, delicious. To describe one of the sensory effects of umami, Shizuko Yamaguchi of the Ajinomoto Company uses the word *mouthfulness*, which, although a mouthful, says exactly what it means. Umami makes the flavor of what's in your mouth feel fuller, deeper. It fills your mouth with more of the flavor of the food it accompanies. Some people refer to this mouth-filling sensation as *round*, as in *That Parmesan cheese has a nice, round flavor.*

Research has proved that humans prefer soup with more glutamate over soup with less, all other tastes and flavors being equal. Other research has shown that adding umami to foods (via MSG in that particular experiment) can make them more palatable to frail, elderly patients who are at a below-average weight, and help them put on weight. It increased the flow of saliva, which is compromised in many elderly people and is important for getting the full flavor from foods. And it somehow also improved immune function, possibly by making the food taste better and making the eaters happier and, as a result, healthier.

MSG can make people eat more, yet the Japanese, who have the lowest body mass index (a measurement of weight and height that indicates health) among rich nations, eat two to three times as much monosodium glutamate, inosinate, and guanylate as Americans. They unabashedly add it to foods. Yet only 3 percent of the Japanese population is classified as obese. Thirty-four percent of Americans are obese. Perhaps the mouthfulness that the Japanese get from their umami-rich diet allows them to eat less without sacrificing sensory satisfaction.

Crave-Worthiness

"It really clicked with me about how umami was probably the taste that you were wanting when you craved certain foods," said Adam Fleischman, founder of Umami Burger in Los Angeles.

"In American foods, pizza and burgers were the ones that were the most crave-worthy. And they seemed to be the ones that had the most umami in them, with the most perfect balance of umami flavors," he said, giving me background on how he decided to open a restaurant focused on umami.

He's absolutely right. Pepperoni pizza is one of the most umami-laden foods on Earth. There's umami in the sauce cooked from ripened tomatoes. There's (a little) umami in the mozzarella cheese, and there's (a lot of) umami in the cured pepperoni. Douse a slice with aged Parmesan and you're gilding the umami lily.

"I wanted to pack as much umami as I could into a burger," said Fleischman. "So I researched all the foods that had the most umami and just figured out a way to get the most umami into a burger."

The chain's signature menu item is the Umami Burger, made from a patty of fresh-ground flap beef topped with grilled shiitake mushrooms, roasted tomato, caramelized onions, house-made ketchup, and a brilliant jolt of unexpected crunch in the form of a pan-fried Parmesan cheese crisp. Fleischman calls the burger "umami × 6." What Fleischman is referring to is the way that different umami compounds can synergistically enhance each other.

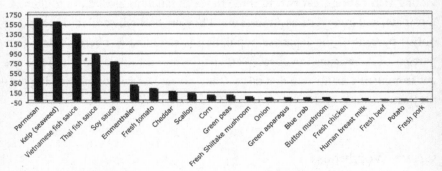

Level of Naturally Occurring Umami

Free glutamic acid (mg/100 g)

Source: K. Ninomiya, "Natural Occurrence," *Food Review International* 14: 177–212.

When I was presented with my Umami Burger, it arrived on a large oval plate, with no garnish or side dish, an entree presentation that demands that the burger be the center of your attention. And, indeed, it was a rich, huge sensory experience. Not a huge burger, though. In fact, I was somewhat surprised at the size of the sandwich. I have had much bigger burgers in my dining career and I've had much richer. I can remember struggling to eat the second half of my burger at Minetta Tavern in New York City, a place renowned for its signature blend of ground marbled and aged rib eye, skirt steak, and brisket beef, which weighs in at almost nine ounces before cooking. It is loaded with umami, as the beef is dry-aged before being ground, but it's a monster. Served with savory-sweet caramelized onions, it was one of the many times I wished I had a bigger stomach. Umami Burger's burger, though, at six ounces before cooking, is huge in flavor, not size.

The mouthfulness of umami gives a sensory satisfaction that differs from anything else. One of the main differences between Japanese cuisine and Western cuisines that are built on French technique is that Japanese food doesn't rely as heavily on fat to carry flavor. Instead, it gives that job to umami. Japanese chefs employ all manner of umami infusion in their cooking. In addition to soy sauce, they use a type of seaweed called kombu to make dashi broth, which serves as the basis for dishes such as miso soup and ramen. Kombu is probably the most umami-rich ingredient in the sea. Ounce for ounce, it's almost on a par with Parmesan, the most umami-rich ingredient on land.

Creating Hugeness of Flavor with Umami

Jon Shook and Vinny Dotolo cook guys' food at their restaurant, Animal, in Los Angeles. Dotolo had described tofu as having the texture of calf brains and joked that he considered putting it on the menu at Animal—tofu with meat. That's when I knew I had to taste these guys' food.

Amid the pig tails, goose liver, myriad sausages, and pork belly on Animal's menu sits a curious fish dish, the hamachi tostada. It sounds innocent enough. On top of a crisp tortilla saucer sits a vertical tangle of fresh cabbage, herbs, onion, and fried shallot. After eating three or four plates of meat, Roger, my sister, and I sighed with relief at the arrival of something crunchy and green. But upon first bite, any thoughts of austere fish swam out of our mind. This unassuming dish rocked our world with a hugeness of flavor that came, unmistakably, from umami.

Dotolo spoke with me about this dish, which he and Shook had originally created for an evening event. "What would be something that would be flavorful but still be somewhat light?" thought Dotolo as he considered but rejected pork belly, given the late hour when it would be served. They didn't want to do red meat, since that was expected of them. The hamachi dish pulled together fresh white fish, a crisp fried tortilla, and crunchy vegetables with a unifying fish sauce vinaigrette. It is a gorgeous example of how an umami-rich ingredient—Vietnamese fish sauce, in this case—can unify disparate ingredients into a coherent whole that doesn't necessarily have to be rich or fatty to deliver huge flavor.

To make fish sauce, tiny fish are salted heavily, sweetened with sugar, and put aside to ferment for two months to two years. During this time, the fish emits an odor so strong it's embarrassing. During a boat tour in 1997 along the Chao Phraya River in Bangkok, Thailand, I first got a whiff of something that made me look at my fellow passengers with suspicion. As it got stronger, the tour guide pointed out the fish sauce manufacturing facility located riverside, by which point we were muzzling our mouths and noses with anything we could get our hands on to stop the stench from reaching our olfactory system.

When fish sauce is fully fermented, refined, and bottled, it retains some of its odorific quality. But it's not the fishy smell that makes chefs reach for it when they're cooking. It's the ridiculously high levels of glutamates that are formed in the fermentation process. Chefs usually add it in such subtle quantities that you'd never know a soup, stew, or sauce contained a fishy-smelling ingredient. All you'd notice is that it was full, round, savory, and delicious.

Umami Moms

In Japan, soy sauce, miso, and dashi go into many sauces. Each of these ingredients is a potent source of free glutamates. In Southeast Asia, fish sauce gives depth to many dishes. In China, hoisin, oyster, fermented black bean, and mushroom sauces are common. They're also all full of glutamate.

Every student who goes to culinary school learns early how to make a stock and, eventually, how to make the mother sauces: béchamel, espagnole, velouté, hollandaise, and tomato. Mother sauces are the building blocks for much of classic French cooking, and it's no coincidence that three of them derive their flavor from umami. Velouté and espagnole start with stocks made from simmering bones, a process that frees glutamates from the animal protein. Tomato sauce is, of course, loaded with umami that occurs naturally in the fruit and is concentrated and further developed in the cooking process. Even béchamel, which we think of today as a white sauce made from milk and butter, originally called for veal, according to the recipe of Escoffier, the French chef who modernized cooking, wrote *Le Guide Culinaire*, and in the process became one of the first celebrity chefs in 1907.

Glutamic acid is the predominant free amino acid in tomatoes, which also include IMP and GMP. As tomatoes ripen, the glutamic acid content—and hence umami—increases, as it does when they're cooked, dried, or heat-processed. So it should be no surprise that three-quarters of the tomatoes eaten by Americans are in some way processed: cooked; canned; turned into pasta sauce, salsa, juice, or ketchup, all of which are chock-full of free glutamates. In fact, many of the foods we eat would be only mildly appealing without the punch of a tomato garnish. French fries are, for example, a vehicle for delivering ketchup—and hence umami.

Taste Two-fer

If you want to get more flavor from your food, you could make one simple switch: replace some of the added salt in your diet with monosodium glutamate. Dare I even suggest this? It's not as crazy as it sounds. Monosodium glutamate contains that same sodium that performs superheroism in foods. But MSG has one benefit over salt: it also contains umami. For the same amount of sodium, you get two tastes in one.

Of course, no one feels good about sprinkling white powder on food, regardless of what it is. The more natural way to work MSG into your food would be to follow Animal's and Umami Burger's lead. Use glutamate-rich ingredients. Instead of water, use chicken or beef broth. Instead of salt, use soy sauce or fish sauce. Garnish just about anything with cooked mushrooms or tomatoes or aged cheese and enjoy a taste two-fer. Umami brings together the ingredients in a dish by adding something akin to time and patience. Keep that in mind when you're trying to combine flavors. Or keep them apart.

Without Balance, Umami Becomes Oooh, Flabby

Just as a dish that's too sweet will taste out of balance, so will one with too much umami. Umami gives such full flavor that overuse of it can be fatiguing. People will generally use the term *intense* to describe a sauce, for example, that's been reduced so much that the glutamates are concentrated into too little volume. If you experience an entree, soup, or sauce that's full of intense flavor yet you find yourself wanting to squeeze a lemon on it, it's likely that the umami is too high. Acid is often the only thing that will tame excess umami. That's why the slices of pickled ginger are there at the sushi bar, and a pickle appears on most of the burgers and next to the sandwiches in the United States.

When you're cooking or tasting food, form a visual image of the Taste Star to jog your memory. There are five Basic Tastes and they all need to be in balance. Umami is no different. If the umami is too high, the dish will taste muddy, too savory, too mouth-filling. The flavor will be indistinct and unidentifiable—and not in a good way. A general savory umami-ness is fine if that's what you're going for. A chef who puts a soup on the menu and calls it lobster bisque probably wants the lobster flavor ringing out loud and clear. Too much umami in a subtle, albeit savory, dish will overwhelm its signature flavor. You're not paying those extra few dollars for umami; you're paying for the lobster essence. On the other hand, if a dish like soup is lacking umami, you'll know it when you take a bite and crave the *something* that Ikeda, Frankenthaler, and Dotolo knew was there all along.

Umami

Measured by: Free glutamates

Classic Umami Pairing: Umami Times Two
Example: Grilled cheese sandwich and tomato soup
Why it works: The umami of the cheese pairs synergistically with the umami of the tomato soup. The crispy edges of the sandwich are a perfect foil for the smooth, sometimes creamy, soft mouthfeel of the soup.

Classic Umami Pairing: Umami Times Two
Example: French onion soup with Gruyère cheese
Why it works: The umami of the cheese pairs nicely with the umami in the (usually beef) broth of the soup.

Classic Umami Pairing: Umami Times Two
Example: Bacon, lettuce, and tomato sandwich
Why it works: The cool, refreshing umami and sweetness of the tomato balances the salt and umami of the bacon. The lettuce and toast add mild crunch that lets the umami shine through.

Classic Umami Pairing: Umami Times Two
Example: Burgers topped with cheese
Why it works: Just about anything works on top of a burger, but cheese is a natural. The dairy aromas work with the beefiness of the meat. If you use an aged cheese, you can really push the umami as Umami Burger does.

Classic Umami Pairing: Umami + Sour

Examples: Hamburger with dill pickle; sushi (dipped in soy sauce) with pickled ginger

Why it works: Umami can be satiating without some sort of relief. The intense umami of soy sauce (or other savory Japanese classics such as ponzu, teriyaki, and miso) benefits from the cutting acidity of pickles. The two balance each other.

Aromas Associated with Umami:

Meat	Fermented	Chicken	Earth	Gamey
Broth	Chocolate	Beef	Onion	Garlic
Savory	Miso	Soy sauce	Tomato	Cheese
Brewed	Funky	Mushroom	Vegetable	

Taste: Isolating Umami

To do this exercise, you will need to buy the pure form of umami, monosodium glutamate. You can find this under the brand name Ac'cent, usually sold in the spice aisle. Since Ac'cent (like all MSG) contains sodium (that's the monosodium part of it), you'll need to compare it against the same amount of sodium minus the umami. This would be salt.

You can do this exercise with any type of salt, but most salts vary slightly in terms of sodium content. If you use Morton salt, which I'm recommending because of its broad availability, I've done the calculations for you. If you use another salt, you'll need to make sure that you're tasting identical amounts of sodium in each mixture.

YOU WILL NEED
> Measuring spoons
> Table salt (not iodized)
> 2 liquid-cup measures
> Ac'cent flavor enhancer or any other pure form of MSG
> Warm water
> Spoons for tasting
> Saltine crackers and water for cleansing your palate

DIRECTIONS
> 1. Measure ⅛ teaspoon salt into the first cup. Measure ¾ teaspoon Ac'cent into the second cup. Fill each cup with warm water to the ⅔ cup line. Stir both with a clean spoon until the crystals are dissolved.

TASTE
> • Taste the salt water first. It should taste salty, like warm ocean water.
> • Cleanse your palate with crackers and water.
> • Taste the umami water next. You will taste salt, as you did with

the salty water, but there's another taste in this sample. That taste is umami. Notice how it tastes meatier and brothier than the salty water. Also notice that the flavor fills your mouth and lasts a long time. These are the flavor-enhancing properties of umami.

DISCUSS

- Umami (in the form of free glutamates) occurs naturally in foods such as beef, Parmesan cheese, mushrooms, and seaweed.
- Umami is also a critical flavor component of foods such as soup, ketchup, soy sauce, and fish sauce.
- See if you can now pick out the taste of umami when you eat one of these foods.

Taste: A Culinary Umami Tasting

YOU WILL NEED

Naturally brewed soy sauce (Kikkoman or other)

Vegemite

Worcestershire sauce

Ketchup

Fermented fish sauce (Thai or Vietnamese)

Grated Parmesan cheese

Ramekins or small bowls

Spoons for tasting

Saltine crackers and water for cleansing your palate

DIRECTIONS

1. Let the ingredients come to room temperature. When they're at room temperature you'll be able to detect more volatile aromas.
2. Portion out a small amount of each ingredient into a ramekin or bowl.

TASTE

- Observe and sample each ingredient and note the Basic Tastes present in each one.
- Note any aromas or Irritastes.
- Cleanse your palate between samples.
- Now go back and retaste them, concentrating on the savory umami in each.

DISCUSS

- What words would you use to describe umami?
- What do the aromas add to umami?
- Which ingredient do you prefer? Why?

Taste: The Effects of Aging and Roasting on Umami

YOU WILL NEED

Young, fresh cheese (be sure it is not smoked)
An aged cheese (at least 18 months old)
Plastic wrap
Fresh Roma tomatoes, 2 quarters for each person tasting
Fresh cremini mushrooms, 2 halves for each person tasting
Serated knife
Baking sheet
Saltine crackers for cleansing your palate
Water

DIRECTIONS

CHEESE

1. Shave the fresh cheese into thin sheets, ideally about 1 inch square. Cover with plastic wrap.

2. Shave the aged cheese into thin sheets, ideally about 1 inch square. Cover with plastic wrap.

ROMA TOMATO, RAW

3. Cut half of the Roma tomatoes into quarters. Remove the seeds and pulp so that just the flesh remains.

ROMA TOMATO, ROASTED

4. Cut the remaining Roma tomatoes into quarters. Remove the seeds and pulp so that just the flesh remains.
5. Roast on baking sheet, without added fat or seasoning, for 30 minutes at 400°F. Cool.

MUSHROOM, RAW

6. Clean half the mushrooms and cut in half lengthwise.

MUSHROOM, ROASTED

7. Clean the remaining mushrooms and cut in half lengthwise.
8. Roast on baking sheet, without added fat or seasoning, for 30 minutes at 400°F. Cool.

TASTE AND DISCUSS

- Sample the cheeses first: young first, then aged. Notice the pronounced umami character of the aged cheese.
- Cleanse your palate with saltines and water.
- Sample the tomatoes next: fresh first, then roasted. Notice the pronounced umami character of the roasted tomato.
- Cleanse your palate with saltines and water.
- Sample the mushrooms next: fresh first, then roasted. Notice the pronounced umami character of the roasted mushroom.
- Can you isolate the umami taste from the cheese, tomato, and mushroom? It adds the depth of flavor that's common to them.

13

The Sixth Basic Taste—and Other Contenders

When I was new to my job at Mattson, I was dropped into the frozen French fry business like a basket of Tater Tots. Our new client wanted ideas for frozen side dishes to sell to restaurants and, given my background in the food service industry, I was deemed the person for the job. Before the brainstorming began, I had to learn everything there is to know about America's number one side dish.

In my youth, my mother used to serve homemade French fries on rare, coveted occasions, but the quantities she made were always short of our appetites. When I first made them myself I realized why: they require a lot of tedious preparation, cutting, blanching, soaking, frying, and frying again. Because of this, many restaurants buy them cut, partially fried, and frozen. I had no idea how the commercial version of the process worked, so I took a trip up to Caldwell, Idaho, to visit Simplot, the place where frozen French fries were born. In 1967, former potato farmer and eventual magnate J. R. "Jack" Simplot made a handshake deal with Ray Kroc to supply Kroc's burgeoning McDonald's restaurant chain; the rest is crispy golden history.

In one fascinating, awe-inspiring, and informative day, I learned what types of peeling, cooking, and freezing processes my wacky ideas would be subject to, from the time the farmers delivered the dirt-caked spuds until they left Idaho in a freezer truck. After touring the 1940s-era manufacturing plant, I wearily made my way back to the Boise airport. I got suspicious looks from the security guards and more at the ticket counter, along with a couple of noses wrinkled in disdain. Was it my urban clothing? My unusually short hair? I made sure I didn't have green stuff stuck between my front teeth. When I finally boarded the flight and sat down, I thought I would be free of judgmental gazes. At that point I heard a buzz from all around me. *Who has the fries? I didn't know there was a McDonald's in the terminal. Does it smell like McDonald's in here to you? Oh, man, I could really go for some French fries.* I had been getting odd looks because my entire body reeked of fried food.

Foods with fat simply savor better. For as long as we've practiced the art of cooking, humans have known that fat works wonders. In ancient Greece, Aristotle documented the "fatty" taste and characterized it as the opposite of salty, anointing it with salt-like legitimacy in his writings. Fat affects the appearance of foods so profoundly that you can easily tell which of two glasses of milk has more fat simply by picking the one that's whiter and more opaque. You can also distinguish between a lean and fat steak visually, based solely on the amount of white fat marbling.

Fat affects the mouthfeel of foods, making them creamier and allowing them to carry flavors longer. And fat affects aroma, changing the earthy, starchy flavor of a raw potato into the familiar, crave-able smell my seatmates might have termed *McDonald's-y*. Sensory scientists these days think something else might be happening when we eat fat, something that we're not even consciously aware of.

Many researchers are convinced we *taste* fat and, as a result, fat should be considered a Basic Taste. According to Purdue University's Rick Mattes, "Free fatty acids don't turn on the other four senses. There appear to be dedicated receptors to capture those unique fatty acid stimuli." And perhaps most convincingly, he notes, "if you cut the taste nerve in [mice], you block responses to fats so it seems that the message is carried by the gustatory system."

This rocked my entire understanding of fat's contribution to food. Clearly fat has a transformational effect on the texture, appearance, and aroma of food. When potatoes are deep-fried, their texture, appearance, and aroma change in ways that stimulate our reptilian brain and tell us that these are not just any old potatoes: they're dense with calories, which Mother Nature has wired us to love so we'll nourish ourselves (or overnourish ourselves, as we do today). But I hadn't considered that fat could have a unique *taste*, as Mattes is trying to prove.

It turns out that isolating the *taste* of fat from its texture is extremely difficult. You can clamp subjects' noses to eliminate fat's aroma, but taste is a *contact sense*. This means that in order to *taste* something, it has to come in contact with your tongue. As a result, there's no way to *taste* fat without *feeling* it.

To get around this, Mattes has tried to mimic the mouthfeel of fat with replacers such as gums, starches, even Olestra (a fat replacer developed by Procter & Gamble). If he could successfully create a fatty-mouthfeel "match," he could isolate the *taste* of fat from any *texture* it possesses by determining whether testers could *taste* the difference between a real fat and a fake fat sample. But this is not so easily done!

Witness the hundreds of companies that tried to sell fat-free products in the nineties and failed. Even Olestra, which has an incredibly authentic fatty mouthfeel, behaves differently in the mouth than fat does. There's a reason that the ice cream aisle of today contains zero brands of zero-fat ice cream. Fat replacers are a far cry from the real thing. Because it's impossible to create a fake fat texture that perfectly mimics real fat texture, people can *feel* the difference, confounding Mattes's research on taste.

Describing Fattiness in Food
(Note That They All Refer to Mouthfeel, Not Taste.)

Lubricious

Viscous

Greasy

Oily

Creamy

Unctuous

Luscious

Slick

The best argument for a fat taste is that those poor taste-blind mice (with the severed taste nerves) stopped choosing calorie-rich fatty chow when they couldn't taste the difference between it and regular mouse chow. They could still get the pleasurable fatty mouthfeel but without the ability to taste they weren't as enamored with it.

We need fat for nutrition, but if you were to chemically separate the fatty acids from the oil or solid fat, you'd get a substance that tastes horrible to 30 to 45 percent of the population. But even the sensitive fat tasters didn't have any better way to describe it than "yuck." This, says Mattes, is another argument for why we need an ability to taste fat. The icky taste of fatty acids may be akin to the unpleasantness of bitter; both function as warning signals. Tasting a yucky free fatty acid may help communicate to us if the fat is rancid and should not be eaten.

But rancidity is detectable by aroma, too. So why would we need to taste it if we can smell it? Again, we don't have an answer. Our taste for fat is a hot topic for more research, as is our taste for other things such as calcium.

Taste, Misplaced

Salt makes food come alive, but Americans eat twice as much salt as we need. Could our collective craving for salty food perhaps be a misplaced craving for something else? Could we be forgiven for ordering a glass of salty tomato juice to wash down our salty snack mix? (And why, for that matter, do people drink more salty tomato juice at 30,000 feet than at sea level? Think about it. At what elevation did you last order a bloody Mary?)

Monell's Michael Tordoff believes that when we reach for salty foods, we may actually be craving calcium. There are two forms of calcium in the blood, a free form and a bound form. The free calcium is the kind that's important for human nutrition. Eating salt frees the bound calcium and makes it more available to the body. If you eat a low-calcium diet, you'll feel better immediately following a salty meal, having freed some of the bound calcium in your body.

Tordoff's research has shown that calcium-deficient women tend to prefer saltier food than those who consume enough calcium. Tordoff also believes that we need to add calcium to the list of Basic Tastes. His team discovered that mice have two calcium taste genes. We don't yet know if this holds true for humans. "But it seems pretty likely we have the same function," Tordoff says.

Ajinomoto, the Japanese company that introduced the world to powdered MSG, has continued to do extensive research on taste, constantly looking for the next great taste enhancer. They recently found that the calcium-sensing receptor can make the Basic Tastes of sweet, salt, and umami savor more intensely. This could explain why we evolved to have calcium receptors: to make calcium-rich food taste better so we'll eat more of it.

If you want to experience the taste of calcium, you have a few options. You could buy some chalk at an art-supply store, plug your nose, and lick it—that's pure calcium carbonate (one of the many forms of calcium). You can buy calcium capsules at a health food store, open one up, dump out its contents, stick your finger in, and taste the calcium. What will you experience? I like to use the descriptive term for the taste coined by Tordoff: *calci-yummy*.

Tiny Bubbles

Altitude sickness is a strange affliction. It seems to attack its victims randomly; if you are struck by it once—or even if you're not—there's no telling if you'll get it the next time you're at a high elevation. The best thing you can do to avoid the debilitating headaches, nausea, fatigue, and worse symptoms is to be proactive and take acetazolamide, also known as Diamox, considered one of the best preventive measures when used along with acclimation and a healthy constitution.

In the 1980s, Stephen Kelleher took acetazolamide and a six-pack of beer up a mountain to celebrate his summiting. When he got to the top he was disappointed to find that the beer tasted "like dishwater." The drug blocks an enzyme that helps detect carbon dioxide, or CO_2. Kelleher concluded, "Mountaineers must choose to leave either their acetazolamide, or their suds, at home." The beer-drinking doctor decided to call this "the Champagne blues."

Research into the Champagne blues led to the realization that we may *taste* carbonation as much as we feel it. The same taste receptor cells that detect sourness also detect CO_2. The scientists who identified this wrote:

> Although CO_2 activates the sour-sensing cells, it does not simply taste sour to humans. CO_2 (like acid) acts not only on the taste system but also in other (ways), the final percept of carbonation is likely to be a combination of multiple sensory inputs. Nonetheless, the "fizz" and "tingle" of heavily carbonated water is often likened to mild acid stimulation of the tongue, and in some cultures seltzer is even named for its salient sour taste.

They pointed out that seltzer in German is called *sauer Sprudel* or *Sauerwasser*, both translated as "sour water."

When I came upon a product called Fizzy Fruit at a trade show years ago, I admired the innovation of infusing CO_2 gas into fruit the same way it's infused into water to make carbonated drinks. The inventors perhaps believed that by adding a unique sensory element to grapes, for example, they could entice kids to eat fruit in lieu of more hedonically appealing, but less nutritionally beneficial, candy. The first time I tried Fizzy Fruit, however, I recoiled instinctively. It just tasted *wrong*. Later I realized that it reminded me of old orange juice that had begun to ferment. And that may explain why we may have evolved to taste CO_2 so we can detect the fizz of fermentation where it doesn't belong, in spoiled foods that have been around too long. In addition to providing a biological reason for the need to taste carbonation, this may also explain why Fizzy Fruit hasn't exactly overtaken the market for candy.

What Is a Basic Taste?

Recognizing Basic Tastes is somewhat akin to recognizing primary colors. There are many colors in the spectrum, but all are made up of some combination of red, yellow, and blue. When you look at green you don't usually think, *That's a combination of 75.3 percent yellow and 24.7 percent blue.* You simply think it's green. Because you synthesize things you see, vision is considered a synthetic sense: the inputs come together in your mind as one coherent perception instead of as parts.

Our sense of smell is also considered a synthetic sense. When you experience a bunch of aromas, you fuse them all together and they become "a tomato" or "a banana," or "McDonald's." You'd be hard-pressed to pick out the individual aroma compounds, even if you know the smell intimately.

Taste, on the other hand, can be considered an analytic sense because we are able to analyze the Basic Tastes in a mixture. Complex aromas are more difficult to analyze. You can easily say that a tomato is sour and sweet, but you have to be trained to identify its aroma compounds.

There are millions of flavors in the world, made up of a combination of the five Basic Tastes, countless aromas, and an unlimited variety of textures. My food-developer definition of a Basic Taste is that which we detect using our tongue alone—like salt—without the benefit of any other sense. The problem of what to consider a Basic Taste arises when scientists are faced with *something* unique that isn't sweet, sour, bitter, salt, or umami. Take water, for example. Certainly water has a taste. When you remove all of the minerals that contribute flavor to waters such as Evian, you're left with a liquid that isn't sweet, sour, bitter, salt, or umami. So what exactly is the taste of water? If we don't have a word for it, but we can detect it with our sense of taste, should we consider it a Basic Taste?

Another definition of what constitutes a Basic Taste is any taste for which we have a receptor on our tongue. With this as part of his definition, Monell's Paul Breslin says, "There might be around twenty qualities of taste but I am most comfortable saying that there's five."

Twenty Basic Tastes? While this may sound preposterous, the field is still new enough that we simply can't say for sure there are only five Basic Tastes. Some people argue that the effort to anoint Basic Tastes is frivolous, but I disagree. The five points of the Taste Star help me think about food in a structured way. I use it as a mnemonic device to make foods taste better. Most people can detect five tastes and I hope my explanations of them help when you're eating or cooking. The Taste Star is a useful concept outside the scientific debate, which is where most of us live and eat.

Top Contenders for Induction into the Basic Taste Hall of Fame

Fully Accepted by Scientific Community	Good Evidence for But Not Full Acceptance Of	Data Are Thin (to Date) But Has Some Advocates	
Sweet	Umami*	Kokumi	Pyrophosphate
Sour	Fat	Water	Lysine
Bitter	Calcium	Metallic	Polycose
Salt	Carbonation	Starch	Hydroxide
		Electric	Soapy
		Mineral	Protein

*I believe umami belongs in the "Accepted" column, but because there are enough scientists who disagree with me, I feel compelled to include it here.

Part Three
Putting It All Together

14

Your Brain on Food

Imagine a barbecue at your best friends' house. The sound of kids playing, the low thrum of Bruce Springsteen on the radio, the *spffft* of beer bottles being twisted open. Now imagine that someone places on the picnic table a platter stacked high with racks of baby back ribs, slathered with barbecue sauce or a dry spice rub—choose your style—fresh out of the mesquite cooker, crusty along the edges, falling-off-the-bone tender, smoky steam wafting off of them.

If you're craving ribs right now, you're human. And if I now present you with a huge platter piled high with individual ribs, already divided into singles for you, it's likely you might eat *more* than you normally would had I not put the image of them in your head and sparked a craving for ribs.

Now imagine taking a rib from that platter and eating it, slowly, then reaching for another one and imagine eating it, and so on and so on—one at a time—until you've imagined eating ten of those delicious, sticky, messy ribs. If I now offer you that real platter of ribs, it's likely you'll eat less than you normally would. Why? Because the simple act of imagining eating a food can reduce your subsequent actual consumption of that food. This is the concept of sensory-specific satiety, whereby each additional bite of a food tastes less and less good. With each bite you are less and less motivated to keep eating it, until . . . you . . . simply . . . stop . . . eating it. This happens even if the eating is happening only in your head. Yes: sensory-specific satiety can occur without the use of your senses.

The experiment that proved you could inhibit actual eating by first imagining eating was conducted by having real people eat real food: M&M's and cubes of cheese. The fact that this works is even more interesting when you consider what's going on in the brain.

At first it seems counterintuitive. We are built to love eating so that we'll nourish ourselves, procreate, and ensure the continuation of the species. It's a nice little system we've got going. Could it be true that the brain will allow us to be satiated with food that doesn't even exist, much less pass our lips? The answer lies in a big white piece of machinery that allows us to see inside the brain, the fMRI (functional magnetic resonance imaging) machine, which resembles an airline fuselage.

The MRI part is what allows us to look at the anatomy of the brain with a futuristic type of camera, by taking an image of it. The *f* preceding this acronym means that the person whose brain is being scanned is functioning, performing some sort of task while inside the camera. The brain image is captured in slices, over time, as the task is being done. The resulting image is the closest thing we have to watching the brain in action. This is how we could see the phenomenon of imagined eating actually working. When you read the fMRI of someone imagining smelling ribs, for example, the image looks very similar to what you'd see if the person actually smelled the ribs. The same holds true for taste: imagining tasting ribs lights up the same regions of your brain that actually tasting them would light up. Of course, when the researchers did the experiments to prove this, they didn't use ribs. They had to pipe tastes into the mouth with a tube and smells into the nose with an olfactometer. You can't eat real food in a normal way in an fMRI because if your head moves when you are eating a rib, the resulting image will be blurry—just like what happens with a regular camera. There's also the problem of getting the rib to your lips without moving and without getting barbecue sauce all over the million-dollar piece of research equipment. For these reasons, brain scans of people experiencing food to date have been somewhat limited to liquid tastes. Even so, they're an ordeal to conduct.

Roger and I visited PhD Joy Hirsch at the fMRI center at Columbia University, where she is a professor and director of the Program for Imaging and Cognitive Sciences. I wanted to understand the science behind how fMRIs worked and, while there, have our brains scanned while we experienced food. She agreed to do scans of us looking at pictures of food. This would be interesting. Roger and I are both hypertasters but we make such different choices when it comes to eating that I thought there would surely be a neurological explanation.

We each chose photographs of food that we loved as well as those we hated. My loves: tomatoes, crusty bread (I knew an image of Tartine Bakery's sourdough would light me up like a Christmas tree), Maryland steamed crabs, sweet potato fries, and Champagne. Roger's were steak, steak, and steak, as well as baby back ribs, ice cream, deviled eggs, burgers, mashed potatoes, and strawberries.

Roger went first, and then it was my turn. I was instructed to lie flat on a platform while the technicians positioned my head in place so they could get a good image. Then someone hit a button and I was conveyed into the fuselage like a duffel bag on a belt.

Meanwhile, Roger sat inside a glass-walled booth talking to the technician who operated the fMRI machine.

"What are you doing?" he asked as the man adjusted some settings on his computer screen.

"Resizing the image field to fit her brain," he said to Roger. I imagine Roger's eyes widening at this.

"Really?" he said. "Whose brain is bigger?"

"Hers."

Roger later told me that my brain was bigger, also having to confess that, yes, he'd asked. Bless his heart.

Of course there is no correlation between human brain size and intellect, I graciously reminded him.

Above me inside the fMRI was a mirror cleverly positioned so that I could see a computer screen while on my back. A technician told me to focus on the screen, and then a slide show of my favorite foods flashed before my eyes, one at a time, slowly enough that they would register in my brain. Then we stopped, the technicians changed the slide show, and I saw my most-hated foods (eggs, pine nuts, liver, and so on).

According to Hirsch, we both showed robust activity in the fusiform gyrus, an area that is activated when bird-watchers look at birds, or car lovers look at cars. She told us that the data indicated we both had a strong interest, and perhaps even an expertise, in food. The rest of the slides showed one huge difference between the two of us. Roger showed much more robust activation when looking at pictures of his favorite foods.

"He's got an emotional attachment to these foods that he likes," said Hirsch. "This suggests to me that what drives his love of food is really a *love* of food. He's driven by his likes, not his dislikes."

I, on the other hand, showed the brain of a restrained eater. I eat what I like and allow myself an occasional indulgence, but I definitely watch how much and what I eat. I restrain my diet. My brain, compared with Roger's, showed much less activity when I looked at the food I loved, and more activity when I looked at the food I dislike (versus his brain when he looked at food he dislikes).

"You seem to be more controlled, more emotionally involved with foods you *don't* like," Hirsch said.

This crushed me. Does Roger live in a happier world than I do simply by not putting any restraints on his diet? Does having a guilt-ridden psyche rob me of some of the joy that Roger experiences daily? And then I wonder whether his HyperTaster tongue has anything to do with this. How much more sensory input does he get than me and, more important, does this translate into more pleasure? Does he get more out of the food he loves than I do?

Hirsch was careful not to read too much into the data.

"It suggests that the individual differences in our brain responses are key to our individual personality differences. It fascinates me that one can look at brain patterns and make inferences in behavioral patterns," she told me before revealing one last truth. "Your brain's a little smaller than Roger's," she said.

"Really?" I asked.

"Female brains usually are smaller than male brains. It's quite proportional to body size," she said.

I should have slipped that fMRI technician a twenty for lying so convincingly to Roger.

Your Brain on Chocolate

Think about the most rewarding food you could eat. For many, this would be chocolate. Chocolate lights up the same part of the brain—the reward center—that is activated by gambling and cocaine. Dr. Hans Breiter of Massachusetts General Hospital writes, "The same neural circuitry is involved in the highs and lows of winning money, abusing drugs, or anticipating a gastronomical treat."

This is all due to dopamine. When you eat chocolate—or throw dice at a craps table or use drugs—dopamine floods your brain. It's what makes you respond the way you do to pleasurable activities like eating. Recent research on the brain's reaction to food has led to a number of theories about why some people overeat and others don't.

One theory says that when the dopamine system works, the result is a very simple cause and effect. Treat: reward. Treat: reward. If you are lean, it's likely that this system is functioning properly. But someone deficient in dopamine receptors may not get a proper reward signal from eating chocolate. It might go something like this—Treat: almost-reward. Treat: almost-reward. When the reward doesn't feel complete, this person may keep going back for more of the chocolate, trying desperately to get a full-fledged reward. Some obese people lack dopamine receptors, so this is one theory explaining why they overeat.

Another theory says that obese people get *more* pleasure from eating than lean people do. In this model, their system rewards them handsomely for delivering a treat, so that they are more motivated to eat the chocolate (or any food) again and again. A third theory says that obese people may *expect* a greater reward, and this expectation keeps them eating long after they are full.

This learning just barely scratches the surface of the brain's role in why some people tend to gain weight and others do not. It's a burgeoning area of science, given the enormous societal costs of obesity and overweight. But it's just now coming together with research on the chemical senses. Our brain is so damn smart, so damn complicated, and so damn inscrutable that we know very little about how it processes most eating.

We know that the Basic Taste sweet shows more action in the brain than the other Basic Tastes, perhaps because it's so important for us to ingest calories. We know that smell lights up more widely dispersed areas of the brain than taste. This makes logical sense since there are an almost unlimited number of smells and only five Basic Tastes.

We know that the sense of taste is more entwined with the sensation of being full and with eating disorders than smell is. Monell's Johan Lundström writes, "The sense of taste might be a good stepping stone, by itself or in conjunction with other senses, in our struggle to understand the explosion in obesity rates and eating disorders."

If we know little about the senses of taste and smell, we know even less about how the sense of touch works in our brain. These days, scientists are most intrigued by the integration of the three senses that combine to form flavor. When we put a barbecued rib in our mouth, we don't experience it separately as the taste, the smell, and the texture. Instead we integrate the information into one perception: the flavor of the rib. What that flavor looks like and where it happens in the brain are still a bit of a mystery. Add vision and hearing to the equation and you're really pushing the boundaries of current scientific knowledge. Lundström is working on a study that will use the fMRI technique to map the brain while a subject is simultaneously tasting, smelling, feeling, hearing, and watching a stimulus food. Of course it won't be exactly the same as eating the food normally, since the subject will be lying in a tomb-like tunnel, with an olfactometer up his nose and a taste tube in his mouth, but it will be as close as we can get within the limits of today's neurogastronomy.

The Expert Eater Brain

One study mapped the brains of expert wine tasters. It showed that the way their brains responded to wine was tremendously different from the brains of novices. There was "higher cognitive processing" of the wine while it was in the mouths of the experts, as well as after they swallowed and experienced mouth-smelling. Experts' practice in the course of their professional wine careers (and ostensibly at the dinner table and elsewhere) made more areas in their brain light up, specifically those in the neocortex.

Not only do tasting experts use more areas of the brain, they can enjoy the processing better than novices. Says Baba Shiv, professor of marketing at Stanford University, "An expert is a person who knows how to derive maximum pleasure from the consumption experience."

Terrible Tastes Taste Terrible

Roger and I were seated at the bar having lunch at a lovely bistro in Healdsburg in the Sonoma wine country. Our grilled asparagus arrived and we fought over the pencil-thin spears: grilled just right, tarted up with freshly squeezed lemon juice, and finished with aged Parmesan cheese. It was a taste of summer, sunny with sweet, sour, bitter, salt, and umami breezes. We enjoyed our leisurely wine-country lunch with requisite glasses of the local agricultural product. My choice, a flinty pinot gris; Roger's, a soft pinot noir. My iceberg wedge salad was crowned with a latticework of perfectly crisp strips of bacon, chunks of funky blue cheese, and a zingy buttermilk dressing. Roger's lamb burger was served with matchstick frites. Both dishes were so good we cleaned our plates down to the last crispy potato crumb. It was a near-perfect meal.

As we were settling our check, I reached for one last swig of water before we headed out the door to begin the dreaded trek back to the city: the event that symbolizes for us the ending of the weekend and the start of the next week. I took a deep drink and thought, *Those damn wine country chefs are always putting fresh herbs in everything. Really, who wants herbs in ice water, for god's sake?* As I brought my hand up to get the twiggy thing out of my teeth, I realized I had been drinking water throughout the meal and hadn't noticed a sprig of herb until now. I froze, slowly pulled out a wriggling spider, and screamed.

I didn't taste the insect, as I was preoccupied with its spriggy-herb texture. When it turned out to be a live creature, the dissonance between what I thought I had in my mouth and the wriggling thing I was holding in my hand was so great that I leaped up and threw the spider onto the bar. The busboy clearing our place settings looked at me as if I had, well, as if I had just pulled a spider out of my mouth and thrown it onto his bar.

That week I was scheduled to meet with Paul Rozin, a professor at the University of Pennsylvania who has done extensive research into why things disgust us. When I told Rozin my spider tale, he clapped his hands and shouted, "Oh, that's a wonderful story!" He then countered with his own.

While he and his then-wife were traveling in Europe, they were invited to dinner at a colleague's house in Switzerland. The man's wife brought a casserole to the table and spooned out a serving for each guest. Paul quickly noticed and unthinkingly blurted out, "There's a long blond hair in mine," clearly associating the strand with the blond hostess. As soon as he realized his mistake, he knew exactly what he had to do to remedy the social gaffe. He promptly pulled the hair out of his food and as his fellow diners watched—since all eyes were now on him—put it in his mouth and swallowed it.

At the University of Pennsylvania since he left Harvard in 1961 with two degrees in psychology, Rozin pursues research across many diverse areas, including food preferences, food appetites, and food beliefs.

I asked Rozin how food can be a source of pleasure one moment and the next moment cause us to gag. The ultimate appetite-suppressing act, according to my research, is seeing someone else chewing with his mouth open, spitting out food while talking, and in general, being a sloppy eater. Why should this disgust us so?

"We live in a disgusting world," he began. "The inside of our body is disgusting. Anything that calls attention to that—that's really upsetting. It's calling attention to something that's happening all the time, to the icky, mushy, wet mass we create when we put food in our mouth, chew, and prepare it for swallowing." He continued:

> Adult eating involves an incredible virtuosity because you are sitting, facing another person, stuffing food into your mouth, making it disgusting in your mouth, and talking through the same hole—at the same time—while looking at the other person, yet you are not presenting them with this disgusting spectacle. That is, you've learned to talk, breathe, and chew this thing without exposing the other person to it. And if they see it for some reason, they're really put off.
>
> We're playing this very delicate game when we eat. We've got this disgusting thing right on the other side of our face. You have a disgusting thing right on the other side of your colon, but you're not at risk of exposing that.

Disgusting things don't need to have disgusting sensory properties. For example, until I knew what I was eating, I thought the spider in my mouth was an herb, which is generally viewed as having positive sensory properties. And the hair that Rozin put in his mouth and swallowed probably had so little flavor he wouldn't have been able to describe it. Yet both of them are disgusting as defined by our culture.

As we saw in the chapter on smell, babies are born without innate smell preferences or aversions (except to aromas that also have a tactile burn, such as ammonia). One-year-olds are completely unfazed by the smell of a full diaper and, until they see mommy's scrunched-up face, don't associate the smell of feces with disgust. It's the same with food: we learn which foods are disgusting not from trying them but from what we tell each other.

The word *disgust* shares a root with the term for taste, *gustation*. Disgust literally means "bad taste." The face you make when you are disgusted is generally the same as the face you'd make if you were trying to get something out of your mouth. The most violent feeling accompanying disgust is nausea, which is also a way of getting food out of your body. Says Rozin, "Disgust originated as a rejection response to bad tastes, and then evolved into a much more abstract and ideational emotion." We in the food business can take pride in the fact that food culture drives popular culture . . . when it comes to disgust.

The most oft-cited disgusting things are human and animal waste products. Rozin writes, "There is widespread historical and cultural evidence for aversion and disgust to virtually all body products, including feces, vomit, urine, and blood (especially menstrual blood)." And feces, says Rozin, "are nothing more than processed food."

Meat Eatin'

We are animals and we consume animals. To reconcile this near-cannibalistic behavior, we have developed a way of distancing ourselves from the fact that we eat dead, chopped-up, heated carcasses. We come up with euphemistic names such as pork instead of pig, beef instead of cow, sweetbreads instead of thymus glands, and Rocky Mountain oysters instead of calf testicles. While we humans are probably the greatest omnivores on Earth, even we don't eat other carnivores. Why? "Other meat eaters don't taste great . . . the meat just doesn't taste good," says Dereck Joubert, wildlife filmmaker and National Geographic Explorer-in-Residence, talking about the big cats in his film *The Last Lions*. Apparently cats don't eat other carnivores because they don't like the flavor. This is, ostensibly, one of the reasons that we don't eat cats. The fact that we harbor miniature, tamer versions of cats as pets also can't hurt in keeping them off our dinner plates.

There are reasons to shun many disgusting foods, but even reason can be trumped by cultural disgust. Some cultures eat insects but in America we do not. It would stand to reason that we'd be afraid to eat cockroaches because they crawl around on dirty things and eat all manner of disgusting stuff. But what if we farmed a pen of cockroaches, fed them nothing but organic vegetables for a month to clean them out, and sterilized them before offering them up on a plate (lightly breaded and pan-fried, I'd suggest, with a sweet chili dipping sauce)? According to the results of a 1986 study conducted by Paul Rozin, even this would be rejected. In fact, my friend Chris and I did something similar.

After learning that domestic yard snails are the same as those raised specifically for eating, we collected a few dozen from my backyard garden. We "farmed" them for a few months in a pen by feeding them vegetables, and then served them up to a dinner party of friends, slathered in butter and garlic. Even our assurances of their organic, hyperlocal (right here!) provenance could not get some of our guests over the psychological hump of eating backyard snails. I mean escargots.

Rozin has also proved that North Americans are reluctant to eat a mound of what looks like fresh dog poop, even after they're told it's made from chocolate fudge. We know that fudge is yummy, but our acculturation to the disgustingness of dog poop simply overwhelms logic. Lucien Malson studied children who had grown up in the wild without human contact of any kind. In *Wolf Children*, he wrote about what types of adults these unacculturated kids became. Without parents or friends to teach them, they did not show any signs of disgust, further proving that culture determines what nauseates us. One culture's Époisses cheese is disgusting stinky sneakers to another; one culture's *natto* is another's disgusting slimy snot.

Bureau of Alcohol, Tobacco, and Fiery Chiles

The fact that humans eat—and enjoy—chile peppers defies explanation. If an extraterrestrial being landed on Earth tomorrow, how would you defend the practice? Imagine the conversation:

> E.T.: What are you eating?
> You: Chile peppers.
> E.T.: What are they?
> You: They're vegetables that induce pain.
> E.T.: What kind of pain?
> You: They burn. Some burn more than others.
> E.T.: Is everyone on this planet as stupid as you?
> You: No, I actually enjoy the pain!

We are the only species on Earth that seems to enjoy the pain response caused by capsaicin, the active ingredient in chiles. (Some types of animals, like birds, don't feel pain from eating it.) Paul Rozin has also done research in this area. He conducted an experiment on rats in which he fed them a spicy food diet instead of their normal rat chow. Eventually the rats learned to tolerate the spicy food. They became desensitized to it. But they never really liked it. After they were used to it, Rozin gave the rats a choice: the spicy food or their former regular chow. They chose the regular. Creatures can learn to tolerate heat but they can't be taught to enjoy it. The love of spicy chiles is distinctly human, but even humans have to be born with the preference.

Human babies don't like the irritaste of chiles and, up until the age of two, will reject it. In chile-eating cultures, a mother may even apply a bit of chile to her breast if she's trying to wean her baby from nursing. But in those same chile-eating households, when the kids are about four years of age, some develop a liking for hot peppers after repeated exposure. Others never get there, even if they are subjected to social pressure from their families, which ranges from simply watching their parents eat chiles to having their parents offer them passively ("Try this salsa, sweetie"), to chiles in home cooking or being exposed to them in restaurants. Kids are desperate to be seen as grown-ups, so that's all it takes with some of them. Being a HyperTaster correlates with the ability to detect heat or pain, since each taste bud is surrounded by pain detection nerves. The kids who never learn to like spicy foods are most likely HyperTasters, with so many sensitive fibers on their tongues that chiles—like many bitter foods—are simply too intense for them.

Some substances require a little bit more of a push. Whereas we can be lightly goaded into liking chiles by our family, other initially icky tastes require a different form of peer pressure, such as alcohol (which burns and has a bitter taste) and tobacco (which stings and tastes bitter). Neither is delicious in a traditional sense, but both offer a reward that some people like and, as a result, become accustomed to the taste. Remember the first time you tasted beer? I'm guessing you didn't love it at first sip. If you did, you're likely a Tolerant Taster. If you don't drink alcohol at all because of its bitter taste (not because of health, moral, or religious reasons), you're likely a HyperTaster.

There's some evidence that consuming chiles is an example of benign masochism: seeking a sensation that on the surface produces discomfort. Other examples include watching a tearjerker movie, riding a scary roller coaster, and gambling. People who enjoy these sensations are slightly more likely to enjoy the pain of chiles.

Perspective

When Steve Gundrum, CEO of Mattson, wants to eat Indian food, nothing will deter him. If something or someone threatens to do so, he simply employs his creative genius in a unique way. This happened when his elderly Aunt Tessie came to visit. She was fairly set in her ways and she absolutely, positively, did not want to go out for Indian food. Too hot. Too much spice. Too exotic. Steve's cogent argument that not all Indian food was spicy did nothing to dissuade her. Instead, he changed tack, and decided to put her in the right frame of mind.

He told his Aunt Tessie that he was going to take her out for some of the best barbecue chicken she'd ever eat. She agreed to this and even got a bit excited. She loved barbecue chicken. When Steve and his family ushered her through the front door of the Indian restaurant, she immediately knew she'd been duped. And she was dually disappointed: she was really looking forward to eating barbecue chicken.

"Just trust me," Steve said. "I promise you won't regret it."

When the perfectly cooked tandoori chicken legs arrived at the table, Tessie melted. They looked exactly like what she'd been anticipating. And so she ate Indian "barbecue chicken" for the first time in her life and loved it. With a fresh perspective, given to her by someone else.

The Wrong Frame of Mind

My mother and her partner, Bob, are avid restaurant-goers who love to find the little hole-in-the-wall restaurants that their home state, Florida, has in spades. They were thrilled to discover a tiny new Italian restaurant run by an authentically Italian family, as fresh from Italy as burrata cheese. They made reservations and went in with high expectations. Unfortunately it was a very busy night and they were given less-than-desirable seating. When they asked the chef-owner if they could change tables, he told them he didn't have anything else available. When they asked if they could sit at the bar instead, the owner politely asked them to leave.

"Why?" they asked incredulously.

"Because you're not gonna like the food. I'd rather have you come back with a fresh perspective."

This was beyond brilliant. The chef knew instinctively that making people sit where they don't want to is akin to serving them a bowl of uncured olives: it would leave a bitter taste in their mouth. His next move showed the chef to be a shrewd businessman. When my parents decided to stay seated at the drafty table by the door, he sent them a complimentary bottle of wine. Upon reflection later that evening, they realized how right the chef was, in both maneuvers. The key lesson learned from their experience is that if you are unhappy in any way about the dining experience you're about to have, it might be best to get up and leave, sparing yourself the disappointment, rather than having to eat a meal in the wrong frame of mind.

A food you love can become absolutely awful if it's served at the wrong temperature. Consider a cup of hot coffee. The aroma of a freshly ground, properly brewed cup of coffee has almost no equal. The smell alone is enough to wake the senses. But if that cup of coffee sits too long and you pick it up and take a swig and find that it's cool, it can be unpleasant. This is because a number of things change when coffee cools. Its fragrance practically disappears as the aromatic top notes settle down and stop volatilizing—something they only do at hot temperatures. Because cold coffee has less active volatiles, it has less flavor. Some of the bitterness that was masked by the beautiful aroma is now sharply apparent. Chemical changes also occur in coffee as it goes from hot to warm to tepid. So the cold coffee is actually physically different from the hot.

Food Tastes Better When You're Hungry

In my twenties I hiked the Inca Trail, the path to Machu Picchu, located high in the Andes Mountains. My then-boyfriend and I were not prepared for the physical challenge that the hike would entail. From the time we put our feet on the ground in Cuzco at 10,000 feet, he battled altitude sickness. Fortunately, I did not.

On the third day of the hike, we planned to arrive at Machu Picchu. When you are hiking at 13,000 feet, the simple act of breathing is difficult and walking on ancient stone trails exhausting. I had skipped breakfast because of my excitement. We arrived at the ancient, mist-shrouded Incan shrine six strenuous hours later. I was beyond famished and couldn't think about anything but finding the snack bar. There I was on sacred ground and all I could think was I need food. Now.

When I finally got my hands on a bag of Combos—a pretzel snack filled with suspicious cheese-ish filling—I devoured it in a low-blood-sugar frenzy. A snack choice I usually wouldn't make nonetheless satisfied me in a way that nothing else ever has or will. Food simply savors better when you're hungry.

Obviously, when your stomach is empty, it feels good to fill it. The more your body needs something, the more pleasant it feels to ingest it. In addition, when you're hungry, your body is ripe with the hunger hormone ghrelin, which makes you sniff more. The more you sniff, the more aroma molecules you ingest, the more flavor you experience. You get more flavor from food, through your nose, when you're hungry.

Being hungry puts you in the perfect space to really appreciate food. We often use this knowledge to our advantage at Mattson by scheduling prototype-tasting meetings at noon. Even reduced-calorie, reduced-salt food will taste better when you're hungry for lunch.

Rumpled Mints

When you have a bad experience with a flavor, such as peppermint, you can develop a conditioned aversion to it. This is probably the most common cause of disgust in the world of food. Remember the man who was averse to coffee after an embarrassing incident in a store with spilled roasted beans? I developed an aversion to enoki mushrooms having once made myself a lovely salad of them, only to become violently ill moments after finishing it. To this day I avoid these long, thin mushrooms that, to me, signal only the induction of vomiting.

Researchers have successfully conditioned aversions with as little as one exposure. Most food aversions are conditioned by stomach flu, food poisoning, seasickness, altitude sickness, painful breakups, or other bad patches of your life. Your experience with a particular food forever colors how you savor it from that point forward.

The Chubby Hubby Effect

When I was in graduate school, I often had friends over to my apartment for study sessions that diverged into full-on parties. One night someone broke out a joint and passed it around. Then we passed around a pint of Ben & Jerry's Chubby Hubby. From the sound of things, you would have thought we were eating food for the first time in weeks. Someone raved about the brilliance of putting salty-sweet peanut butter pretzels in vanilla malt ice cream. Another person added that the fudge and peanut butter swirled throughout were strokes of genius. Without a doubt, we all agreed, Chubby Hubby was the best thing we'd ever tasted.

The following week I treated myself to a bowl of Chubby Hubby before bed on a Tuesday night. Was it delicious? Of course. Was it the best thing I'd ever tasted? No. The altered reality in which we tasted the ice cream had altered our perception of the product. From that point forward, I dubbed this The Chubby Hubby Effect. I decided to conduct some of my own pseudoscience to prove or disprove this phenomenon. I conducted a lack-of-focus group.

Roger and I hosted eight friends, selected for their open minds and willingness to smoke pot. We set up tasting plates for each attendee. When they arrived—and before they could inebriate themselves—we asked them to sample five foods and fill out a sensory evaluation form for each. The samples included a Southern-fried chicken tender and a wedge of Parrano, a Dutch cheese that is semi-aged and just shy of the umami wallop of aged Gouda. We threw in Belgian endive to represent bitter, an Italian dry riesling as the sour entrant, and a milk chocolate bar from Scharffen Berger for sweetness and mouthfeel. Then we smoked the weed. Roger remained sober throughout to help administer the test and make sure things stayed on course. When he saw that we were all good and stoned, he made us taste the same five foods and fill out the sensory evaluation forms again, an exercise akin to herding cats.

The most articulate comment from the lack-of-focus group explains what we experienced: Being high sharpens the things you love and elevates your experience of them and also dulls some details that may usually influence your experience of the food (ugly plates, lack of aesthetically pleasing presentation) so you can focus on the parts that are important to you.

I experienced the Belgian endive as less bitter and much sweeter when I was high. The wine was less sour, more sweet. Just like everyone else, I liked the chocolate from the outset, but liked it a lot more when high. Sweet tasted fantastic.

The common theme was that getting high intensified one's perception of the five Basic Tastes, but our ability to focus on the subtle nuances of aroma, and thus flavor, was muted. The stoned tasters gave very few well-formed thoughts, scribbled in barely legible handwriting, but their comments focused on how sweet the cheese or chocolate was, or how sour the wine was or was not. Later, one guest e-mailed me: "Bland things are awful when you're high. For instance, when I'm serving a roast chicken and simple salad for dinner, I will not let anyone smoke pot before. As a hostess, you want people in the right space to enjoy your offerings, so you have to be proactive and manage the experience. If you are ordering up a pepperoni pizza, then it's OK to be stoned out of your mind!" A pepperoni pizza is loaded with salt, umami, sour, and fat. It's certainly not subtle.

In fact, dozens of studies have been done on how cannabis affects the perception of taste, most of them double-blind and placebo-controlled. The majority of the studies are done with the active compounds in pot, known as cannabinoids. Rick Mattes of Purdue University College of Consumer and Family Sciences has studied many different routes of delivery of the drug, some of them more obvious than others: oral, intravenous, smoked, sublingual (under the tongue), and rectal suppository (the least likely form to be abused by buzz seekers).

Mattes's work was for medical purposes, to determine whether he could increase a person's appetite using pot. If he could truly prove that this worked— as opposed to its being anecdotally believed to be true—the findings would be beneficial for patients undergoing chemotherapy or radiation treatment, when appetite disappears and nausea prevails.

In two of Mattes's studies, the subjects were given the drug, then allowed to choose from a veritable stoners' paradise: an unlimited supply of Oreo cookies, cupcakes, M&M's, fruit, potato chips, peanuts, cheese, crackers, pickles, yogurt, sour candies, juices, bittersweet chocolate, radishes, walnuts, celery, and raw broccoli, among other munchies. Not surprisingly, sweet foods were chosen more than others. This tracks with my lack-of-focus group, which was drawn to sweet tastes.

Mattes was unable to prove in his study that pot increases appetite, but he's not convinced he did the experiment properly. With hindsight, he thinks he used doses that were simply too strong. While he didn't see the appetite-enhancing effect in his study, he heard anecdotally from his subjects that after they went home and some of the drug wore off, they attacked their snack pantries and refrigerators with fervor. Another interesting aside: Mattes says that smoking is by far the best method for administering medical marijuana, because you can titrate (adjust) the dose on the fly. Once you've eaten pot, there's little you can do to correct the dosage, other than grab a package of Oreos, queue up a Monty Python movie, and wait.

Mattes also conducted a similar study to see if cannabinoids affected the sensory perception of food. Other studies have shown that odor perception is not affected by pot, so he focused on the Basic Tastes but did not find a significant effect of the primary psychoactive component in marijuana on taste intensity.

Other studies on sound, sight, and touch were also unsuccessful in demonstrating that they were enhanced by pot, and Mattes speculates that reported enhancements "may be more attributable to effects of the drug on memory and cognitive processes than to alterations to the sensory systems."

In other words, The Chubby Hubby Effect may actually exist, but only in our heads.

Pleasure Plus Pleasure

When I talked to Baba Shiv, professor of marketing at the Stanford University Graduate School of Business, he told me how pleasurable experiences can have a multiplier effect on one another. As an example, he told me that eating popcorn in front of a movie you're enjoying can enhance the pleasure of the popcorn and a pleasant bowl of popcorn can enhance the movie. This immediately made me think of an "only in San Francisco" example that I've experienced.

Every now and then the wonderfully quirky Castro Theatre in San Francisco runs my all-time favorite musical, *The Sound of Music*, on its historic wide screen. The Castro, in all its gaiety, runs subtitles at the bottom of the screen as encouragement for the audience to sing along with Maria and the Von Trapp children. People take this very seriously. Many even don costumes. (Think of large gay men in nuns' habits and grown women in lederhosen.)

The Castro also happens to serve a killer tub of popcorn. As much as I love this event, perhaps reveling in the silliness and fun of the sing-along *Sound of Music* has enhanced my memory of the theater's popcorn. According to Shiv, "The overall pleasure that the brain would code is going to be greater than if it just had the popcorn without watching the pleasurable movie."

The brain takes all the inherent pleasures of the popcorn—the crunch, the buttery aroma, the saltiness, the fun of popping it into your mouth—and integrates it with the pleasure you're experiencing while you eat it: the freeing feeling of singing at the top of your lungs, the sense of belongingness as 1,600 people sing aloud together, the fun of pretending you're sixteen going on seventeen. Then it adds the two sources of pleasure together. It's a recipe for getting the absolute most enjoyment out of food. The flip side of this is that it's the perfect storm for overeating.

"The experience of pleasure is not unimodal. It is not just coming from the taste buds. It is coming from all the other senses, too," says Shiv.

This holds true for the best meals of your life. It's likely that they were with friends and family, all having a jolly good time, perhaps even celebrating a momentous occasion. All this non-food-related pleasure is added to any pleasure you get from the sensory experience of the meal, thereby making it one big, huge happy memory. Says Shiv, "Some of the input is related to the tasting experience, some of it is unrelated to the tasting experience. The brain doesn't make a difference between the source of the emotion."

Wine Placebos

Baba Shiv's marketing research seeks insights that are relevant to the types of companies Mattson has as clients, and can inform the way they market their products. Shiv separates consumer behavior into two components: wanting and liking. Not only is there a difference in the feelings of wanting and liking a food, these emotions are coded in different parts of the brain.

At the end of the day there's nothing I want more than a glass of wine. I'd like to think I'm far from a wine snob; rather, I'm just persnickety in terms of what I like. My everyday drinking wine is a white blend from Oregon that retails for $16 a bottle. And I'll often join Roger in a glass of his soft pinot noirs, which typically sell between $25 and $40 per bottle. Anything more expensive than this is lost on me. I just don't experience the incremental benefit. But perhaps I would if I used an fMRI machine to track my brain through the process. Shiv has done this with wine drinkers.

He put human test subjects in a scanner and squirted five wines into their mouths one at a time. The consumers were told they'd be sampling different cabernet sauvignons. The conceit was that those running the test would identify the wines by their price point only, instead of assigning them random letters or numbers as is normal research practice. Price was the only information about the wines that they gave the consumers. Of course you can guess that they served the consumers the same wines more than once: at different price points.

As you might also expect, when unsuspecting consumers tasted "$90 cabernet" they gave it higher liking scores than they gave the exact same wine labeled "$10 cabernet." But this is where it gets interesting. When Shiv's team looked at the brains of the tasters, the $90 version of the wine resulted in much more activity in the medial orbitofrontal cortex—the area that codes for pleasure—than the $10 version of the same wine. The brain seemed to be experiencing more pleasure from the higher price point.

Shiv explains how this works. When presented with a higher price point, we expect a higher reward, since the expensive-means-better model has proved true in most other aspects of life (cars, houses, educations, vacations). As a result, when our brain hears "$90 cabernet," it gets ready to receive this high-value reward. With our brain primed and ready, the act of actually tasting the wine fuses with the expectation and we experience a higher reward than we do for a wine with a lower quality expectation. This is nothing more than the placebo effect! When you expect a drug to work on your ailment and—more important—when you want it to work, it oftentimes will. When you expect a wine to savor better, and—more important—when you want it to, it will often do just that.

He tells of another experiment he did with beer. (After this story I decided that I needed to visit his research lab at happy hour.) He proved that people have different reactions to the same beer if it's served two ways: with a twist-off cap or with a cap that requires a bottle opener. Needing to find and use an opener on a bottle of beer requires a little bit more effort. Not much, but enough that you notice it. But it also does something else that's extraordinarily important. It prolongs the wanting. It adds another layer of anticipation, which translates into a higher level of liking. As Shiv says, "It is in the anticipation of pleasure that pleasure itself is provided."

This is a big problem in our society today. We have such incredibly easy access to food. It's usually only a few feet away from us, ready to heat or eat, with no patience required. We do a lot less cooking, outsourcing this task to restaurants, delis, cafeterias, and others, so our wanting of food is easily satisfied. We don't allow our wanting to build up, and so we cut short a source of pleasure: the anticipation of it.

Taste in Space

Imagine the eater's worst nightmare: You are living in captivity for six months without fresh food or water. You cannot eat anything that hasn't been freeze-dried or canned. There is not a piece of fresh fruit available. No cold, cool, fresh milk, creamy cheese, or comforting ice cream. No grilling or baking. And worst of all, your body feels, at best, *off*.

There are people who volunteer for this kind of torture. They're called astronauts.

I was curious about how our sense of taste would be affected when the atmosphere changes. Food doesn't taste as good at 30,000 feet as it does on Earth, but what about a more extreme change? After all, gravity and air are present in a pressurized plane cabin. So I went a bit higher.

I spoke with Michele Perchonok, a PhD in food science, who manages NASA's Advanced Food Technology Project and the former Shuttle Food System. "I have probably one of the very best jobs a food scientist could have," says Perchonok about her job at Johnson Space Center. "I wouldn't want to be anywhere else."

Perchonok's team created the food for astronauts on the recently retired shuttle, as well as those who spend time on the International Space Station, some for stretches as long as six months. Her team's challenges are many. First and foremost, they have to make sure that the astronauts are well fed and well nourished. Astronauts' time is highly valuable when they are in orbit. Each minute some sort of important work or activity is scheduled, and the astronauts must be in peak condition. They cannot afford to be either hungry or stuffed.

Then there's the issue of storage space when people are eating in outer space. There's no room for food detritus, such as the empty jars, bottles, cans, wrappers, skins, rinds, and cores that are generated by normal food preparation. As when camping or sailing: whatever you pack in you have to carry around and ultimately pack it out when you go home.

Most of the hot meals that astronauts eat are processed in the same way that canned foods are, so that they're sterile and don't require refrigeration. But instead of bulky metal cans, the chicken chunks, brown rice, or franks-and-beans come in foil pouches, like premium albacore tuna pouches you might see at the grocery store. The benefit of foil pouches is that they collapse flat and can be folded or rolled up easily to minimize their volume.

The options for preparing food on a space shuttle or space station are limited. There are no electric or gas grills or stoves, because it's difficult and dangerous to use high heat in small, closed environments. In fact, no actual cooking is done in space. The food that's sent up is already fully cooked. The astronauts merely warm or rehydrate it before eating.

Once Perchonok and her team have decided what type of food they are going to send into space and sterilized it in collapsible packages, they have to figure out how to make it taste good. Relying simply on culinary and food science techniques would assume that everything is the same in orbit as on Earth, which is untrue. Let's start with the eater.

When astronauts first arrive in space, it takes some time for their bodies to adjust. Because of the lack of gravity, bodily fluids tend to move around in unusual ways. Fluid flows from the body core and limbs into the head in ways it never does on Earth. You could approximate this if you hung upside down for an extended period of time, but there's no good way for you to simulate zero gravity. The astronauts start to experience what they gently refer to as "Charlie Brown face," meaning that their faces become plump and round from the extra fluids. Besides looking silly, the other effect of this is nasal congestion, which makes eating food about as enjoyable as it is with a head cold. They can experience the five Basic Tastes fairly well, but the aromas are severely muted, so flavor is compromised.

"We get a lot of comments from our crew members saying food just doesn't taste the same in orbit. Most of the time they say it doesn't have enough flavor," says Perchonok. This is because the meals are designed to be moderate in sodium content. The issue is that high salt consumption in space can lead to water retention and other health problems. And when the new superlowsodium formulas that Mattson developed make it to the Space Station, there will be even lower salt, meaning lower flavor, less volatiles that are salted out, and less of that yummy salty taste.

The astronauts eat out of individual pouches, as opposed to putting food on plates. "It's very difficult to transfer food in microgravity (or zero gravity)," says Perchonok, "because it could float away." The other reason for not putting it on plates or in bowls is that using serving ware results in more solid waste to pack in and out, not to mention having to use precious water to wash them.

The third factor is texture. Canned food loses a lot of its freshness cues, the main ones being texture and color. Green beans, corn, and peas that pop when you bite into them when fresh become mushy and soft when canned. The same goes for just about every other ingredient, save perhaps the water chestnut, a canned food formulator's dream ingredient. Without the crisp *snap!* sound of fresh vegetables bursting against the teeth, another sensory input is limited: the auditory kind. And of course the brilliant green color of sweet peas becomes drab military green after canning. When the astronauts get to see the food—which happens only while it's on their fork between scooping it out of the pouch and putting it into their mouths—it is a visual letdown.

Anyone, including crew members, would take less emotional enjoyment from food they have to scoop out of a foil pouch, very carefully, one spoonful at a time, directly into their mouth. It just doesn't give the same satisfaction as cooking, smelling, and tucking into a big bowl of highly textural, rainbow-hued food.

Finally, the astronauts' food is merely warm, not hot, which limits the release of the volatile aromas that higher temperatures produce, muting the flavor further. Because they're in zero gravity, aromas that would normally waft up from the food to the nose may waft, but there is no *up*, per se, so less aroma reaches the olfactory system.

"We send up a lot of condiments," says Perchonok, most of them spicy, because the sensation of heat comes not from the taste or smell systems, but from pain fibers or tactical sensation. This, apparently, doesn't change in space. If and when we meet our first extraterrestrial neighbors, we might want to show up with a bunch of dried jalapeños as opposed to a more traditional peace offering.

The condiments also give the astronauts the illusion of choice, which they severely lack in orbit. Every minute of their valuable time in space is preprogrammed with activities, from when they will sleep to when they will perform experiments to when they will eat.

Other psychological effects come with being 190 nautical miles above Earth. The astronauts are under constant low-level stress, far away from loved ones, and this puts unattainable expectations on the meal: it's the best part of their day. Even if the food were prepared by a James Beard Award—winning chef with the benefit of gravity, no meal could deliver satisfaction, nourishment, control, release, a taste of home, leisure, escape, and everything else someone needs when stepping outside our atmosphere. Space is possibly the worst environment in which to be able to fully experience and enjoy food.

But there's no better view in the universe.

15

How Taste Affects Your Waist

Humans have junk food faces. Daniel Lieberman, Harvard professor and author of *The Evolution of the Human Head*, writes, "There is much circumstantial evidence that jaws and faces do not grow to the same size they used to *precisely* because of our softer, more processed diets."

A chimpanzee can spend half its waking hours chewing because its plant-based diet requires a lot of mastication to break down the cells of plants into something they can swallow and digest. This gives "slow food" a new meaning. Humans, on the other hand, spend less than two hours a day eating meals, and much less time chewing, which takes up a small fraction of mealtime.

Our industrious species has learned how to mill, tenderize, cook, blend, and bake nutritious food that's dense with calories. The majority of what we eat is so soft that we can spend a lot less of our time chewing and a lot more of our time surfing the web. But unless you're chewing food, you're not consciously savoring it. So let's think about this differently. Those lucky chimps! They spend half their day *eating!* Wouldn't it be great if we could spend more waking hours with delicious food in our mouths?

We could, but we'd have to change our diets to food that requires significantly more chewing. Some people have tried, but it's hard to get proper nutrition from plant-based, raw food. There's no need to be so extreme, though. We just need to work a few more jaw-challenging bites into our repertoire.

Horace Fletcher, an overweight nineteenth-century entrepreneur, devised his own diet regimen, the basic tenet of which was to slow down and chew often, lengthily, and repeatedly. He refused to swallow food until every last drop of flavor and every last shred of texture was gone from it. He once chewed a green onion 700 times before he deemed it ready to be swallowed. This style of eating became known as Fletcherizing. Mothers around the United States harped on it, urging their children to chew every mouthful no less than forty-five times before they swallowed. But even at the turn of the twentieth century, people were too busy to masticate every mouthful for three minutes or more, even if they believed it was a good idea. (I don't, by the way.)

Savoring food in the process of all that chewing, Fletcher wrote:

> When one comes to think about it, what sense is there in throwing away a palatable morsel of food when the taste is at its best, or while taste lasts at all, even if the purpose of the meal is merely to contribute to the pleasure of eating? . . . The desire for acquisition, sometimes called greed, impels one to swallow one mouthful of food to take in another, without ever dreaming that the very last contribution of taste to the last remnant of a delicious morsel is like the last flicker of a candle, more brilliant than any of the preceding ones . . .

> Taste, if allowed to serve its full purpose, furnishes its own draught of cheerfulness by means of the very pleasure it distributes, and at the same time it prevents, instead of inducing, gluttony.

Fletcher wrote this in his 1903 book, *The New Glutton or Epicure*, by which date he had lost much of his excess weight and become a health evangelist. Although it was eventually considered a fad diet and faded away, Fletcherizing had its followers, including John D. Rockefeller.

Like many things in life, there may be some benefit to Fletcherizing in moderation. Next time you eat dinner, swap your normal side dish for some raw vegetables. Because you will need to chew more in order to get the vegetables ready to swallow, your meal will last longer, and when you're finished, you'll have eaten fewer calories than if you'd eaten a more highly processed side dish. As you're chewing those carrots, consider how lucky you are that your species has conquered fire and you can choose when to eat raw food. Revel in the natural sweetness of the vegetable, its crisp, almost woody texture, its vivid orange color, and thank the jaw workout for reminding you that you don't have to live on raw food alone. Of course, even if you're eating soft foods, by chewing them—or swishing them around more—you'll get more flavor from them.

Another consequence of our soft food diet is that our teeth no longer fit in our mouths. Because we don't engage in the vigorous chewing that our ancestors did, our jaws don't grow as big as they used to. The result is that our wisdom teeth have nowhere to go, end up impacted, and have to be cut out. If we ate the same fibrous, chewy foods as our ancient ancestors, we'd spend far less money on orthodontics than we do today.

No Smelling, No Belly?

When I was reviewing restaurants I met a woman, Deborah, who lost her sense of smell and proceeded to lose thirty, almost forty pounds. She simply stopped eating. This got me thinking. What if all we had to do was temporarily disable our sense of smell in order to lose weight? What an easy thing to do: clamp your nose shut for six months, lose thirty pounds. I even envisioned a simple but fashionable nose clip, perhaps designed by the likes of Tom Ford, Hello Kitty, or Nike. It could be the inexpensive, reversible alternative to bariatric surgery.

Then I consider my olfactorily challenged friend, Carlo Middione, a professional chef who lost the sense of smell in a car accident. Carlo, like many people who lose their sense of smell, immediately began to *gain* weight. He told me of eating and eating and overeating, hoping in vain that the next elusive bite would be able to satisfy him the way it used to when he had a functioning sense of smell. So which is it? Does knocking out your sense of smell result more often in weight loss or weight gain? At this point, there aren't any definitive studies on this. So I went to someone who works at a Center for Taste and Smell, figuring she'd have the answer.

"The fact is, loss of taste and smell generally doesn't produce weight loss," said Linda Bartoshuk matter-of-factly. People who lose their sense of smell don't lose their hunger-producing mechanisms. In other words, they still get hungry. They eat because they're hungry. The amount and type of food they eat may vary by person, but if all their other bodily systems are working fine, when their stomach empties, they'll feel hungry, just like someone with an intact sense of smell. They may eat *less* because the food doesn't savor as good as it used to, like Deborah. Or they may eat *more* because it doesn't savor as good as it used to, like Carlo. Or their loss may not affect the amount of food they eat. How people handle smell loss is as dependent on the individual as how they handle other major changes in their life.

Of course, there are other circumstances that can cause weight loss, which may occur at the same time that someone loses sense of smell. Oftentimes these other things become the scapegoat for weight gain or loss. For example, chemotherapy can result in a loss of taste or smell as well as debilitating nausea. Which one is responsible for weight loss during cancer treatment is hard to say. It's different for each patient. Let's not forget the emotional trauma of being diagnosed with and treated for cancer, or being injured in a car accident. Emotions spiraling out of control can result in a change in diet. Whether you eat more or less when you're undergoing stress is probably apparent from your body mass index. The point is, we're all different.

Breed a Better Eater

What if we could raise children who would gladly finish all the vegetables on their plate without having to negotiate peace treaties to get them to do so? What if we could raise kids who preferred the flavor of fruit to candy? What if tweens would request broccoli and spinach for their birthday parties in lieu of cake? Could we accomplish this if we understood what drives food preferences? Of course, we don't have the whole picture (yet), but a basic understanding of sensory science can start us down the road.

First of all we need to remember that approximately 25 percent of the population is made up of HyperTasters. These supersensitive individuals are going to be very difficult to train. So let's start with the easier ones: the 75 percent of the population that's made up of Tasters and Tolerant Tasters.

Remember that children start to learn flavor preferences in utero. Researchers at Monell proved this with the carrot experiment. So the first and easiest step is to advocate—strongly—for pregnant and nursing mothers to eat more healthful foods. Yes, this sounds naive and obvious, but it's not that they need to eat *only* healthful foods. They just need to expose their growing kids to those flavors during the critical periods of pregnancy and breast-feeding. Their kids need to be exposed to fish. They need to be exposed to a wide range of bitter vegetables. They need to savor tea, red wine, walnuts, salmon, and other complex-tasting foods that have been proved to provide health benefits.

Once kids are born, we need to nurture them to develop their palates. This means letting them know that food that tastes good doesn't have to be overly sweet, overly fatty, or overly salty. We need to communicate that bitter means health. Cultures that believe bitter foods are healthful tend to eat bitter foods more readily simply because they think these foods signify health.

I believe that part of the reason that lower-income communities suffer more from obesity, hypertension, and diabetes is that they never learn to appreciate bitter tastes. Lower-income households tend to (on average) eat more processed foods. On average, processed foods contain far fewer bitter tastes than fresh. If you grow up in a household where bitter foods are not served, you'll never learn that vegetables can taste good.

If we want to improve the health of children in low-income communities (and all communities, for that matter), there's an easy class we could teach to give kids an appreciation for the flavor of some healthful foods. We could take a page from the supersour candy playbook and make tasting bitter foods a daring, laudable accomplishment. In other words: a game. Who can tolerate radicchio the best? Who makes the fewest faces when eating Brussels sprouts? And who actually likes the flavor of radishes? The more we teach kids that bitter foods are usually telling you that they're healthy, the more information we arm them with. And, of course, the more fun we can make it, the better the lesson will stick with them.

Out of Touch with Hunger

I challenge you to remember the last time you experienced really angry, gnawing hunger pangs that lasted for more than a few minutes. For most people in developed nations this is hard to do. The minute we feel the teensy itch of hunger we satisfy it. We spoil our hunger like a precious newborn, stuffing a metaphorical breast in its mouth each time it threatens to cry. As a result we've lost touch with hunger cues.

When we sit down—or worse: stand or drive—to eat we don't really know if we're eating because we're a little bit hungry, bored, or famished, or just because it's the time of the day when we normally eat. I wanted to see if the scientific community agreed with my perspective, so I called Patricia Pliner, a social psychologist at the University of Toronto. She studies why people eat what they eat as well as why they eat the quantity they do.

I asked her to give me her professional opinion. Does our ridiculously easy access to food from the time we wake up until the time we fall asleep play a bigger role than hunger in what and how much we eat?

"Absolutely. One hundred percent," she confirmed. "I think that—except under pretty extreme circumstances—the amount people eat is dictated by social norms and the presence of food to a much greater extent than by what you might call hunger or satiety. I think those two things are very unimportant in determining how much people eat."

Unimportant? Hunger and fullness are unimportant? We think we eat when we get hungry and stop eating when we're full. But here is someone who has conducted research on hunger, who tells us that this doesn't happen. Pliner then told me of an experiment she did that proved this fact.

First, she brought people into her lab and gave them a set amount of food: a bowl of chicken noodle soup, crackers, a turkey sandwich, and strawberry yogurt with fruit. Exactly 369 calories. Half of the people were fed this amount of food while they stood alone at a counter and ate it in one standing. This scenario was meant to mimic the way we snack. The researchers also primed these subjects with snack-y language. The other half of the participants were taken into a room where each was seated across from another subject at a dining table, while music played in the background. Their food was divided into three courses: a soup starter served with crackers, a sandwich entree, and a yogurt and fruit dessert. It was served on real plates with real silverware. Same caloric content: 369. This scenario was meant to mimic a formal meal setting. Later, when both groups were offered an unlimited amount of pasta, those who had eaten casually standing up ate more than those who had eaten in a more formal meal setting. Remember, both groups had eaten the same amount of food before being offered the pasta.

The takeaway: don't eat standing up! More seriously, the results show that simply changing the way you think about snacking can change the way you eat later. If you are really hungry and feel that you need to eat a substantial snack of, say, a handful of crackers, cheese, and some soup, sit down at a table, use plates and silverware, and convince yourself that this amount of food is a small meal. Then, when you sit down to eat your next meal, pass on the appetizers and soup, or maybe eat a smaller entree, because you've already ingested a small meal in the form of cheese, crackers, and soup. They count against the whole amount of calories you'll eat in a day. But instead of seeing the glass as half empty, consider it in the positive. If you do it right, you can treat yourself to more than three real meals a day! What a luxury. What indulgence!

Snacks—especially substantial snacks—are not free. This insidious self-delusion is part of what has made the Western world fat.

Teach Yourself to Like More Healthful Foods

Let's return to the concept of adaptation: the idea that the flavor of a food or drink is perceived as being weaker and weaker with repeated bites. Put another way, you need more and more of it to be stimulated. Consider that the reverse may also be true.

When I set out to write this book, I also set out to improve my personal nutrition. I am a fanatically healthy eater, subsisting almost entirely on fruits and vegetables for the first two meals of the day plus the dozens of bites I take in the lab at work. I indulge myself at dinner with a restaurant habit that seems to get worse every year as I get busier. I'm lucky that I can afford to buy and eat fresh foods prepared well all the time. But I simply cannot control my sodium intake. This is not just a trivial matter for me. I lead a self-inflicted high-stress lifestyle. My mother and brother suffer from hypertension and my father did, too, before he died of a massive heart attack. When I first started to understand the concept of adaptation (earlier defined as how sensitivity to a stimulus is decreased with each additional tasting), I thought it could be helpful for me *in reverse*, to reduce my salt intake.

Nutritionists preach reducing sodium intake. But it's too radical a change to go from deliciously salty potato chips to the low-sodium (or, God forbid, salt-free) kind. This type of shock to the system causes people to fall off the wagon, as happened to me during tomato season. The trick is to slowly decrease the amount of sodium in your diet until you adapt to the lower level. Then decrease it again. In fact, this is exactly what the packaged food industry is doing. Years ago there were many famously failed attempts at introducing low-sodium versions of foods. The problem is that humans can't readily adapt to a large percentage cut in their sodium. If you're eating regular chicken noodle soup on Monday and you try a low-sodium version on Tuesday, you're going to be overwhelmingly disappointed by how much you miss the salt. It's not our choice to be disappointed by low-sodium foods. It's the way our taste system operates.

Today, however, the food industry has finally absorbed the science and manufacturers have vowed to slowly decrease the sodium in their formulas. They plan to do this slightly and imperceptibly over the course of the next few years. Since consumers have proved that they can't easily adapt to lower-salt products, the food industry is going to do it for them. This is one tactic for resetting the unhealthy norms we've gotten used to in the United States.

The Semisweet Initiative

For the past fifteen years that I've been working in food development, I've been hoping that Coke or Pepsi would call me with an initiative around my pet cause, my own "semisweet initiative." I want to create a soda that is less sweet and less bad for your health. Alas, they haven't. Then I landed a carbonated beverage project with a major grocery store chain (for their own brands, formerly known as "private label" or "generics"). I knew right away that this was my opportunity to pitch my idea.

The food industry's extraordinary ability to reformulate products that are healthier—but taste similar to the unhealthy ones—has done a huge disservice to the population. Let's use the example of a carbonated cola. If we wanted to make a cola that is less bad for consumers, we could start by removing the sugar (or high-fructose corn syrup, but that's another issue). So for years the food industry has been removing the sugar and replacing it with sweeteners that mimic the same hedonic level of sweetness in the sugar version. In fact, we've gotten very, very good at it, but this does nothing to retrain consumers' taste buds to like colas with less sweetness. In fact, it teaches the consumer just the opposite: that beverages are supposed to taste exactly as sweet as they've always tasted. This is the wrong approach.

A better alternative would be to slowly reduce the amount of sugar and simply not replace it in order to create a reformulated drink that tastes *less sweet*. This is the only way we're going to recalibrate consumers' palates for sweetness, exactly as I'm suggesting for salt. The only reason the food industry hasn't been quicker to reformulate food with less salt is that there's no silver bullet for sodium in the way that sucralose or aspartame works for sugar. As you remember from our experience with the NASA meal program, every single flavor in every single product requires rebalancing after you remove salt from the equation. Still, the point is that what you eat on a daily basis sets your norm for what tastes right. But there is no definitive right: you can change your norm.

When I presented the concept of Semisweet Soda, the grocery store chain was very receptive to it. I wrote it up, and while we tested the idea with consumers, Silvina Dejter, a Mattson product developer with a fantastic palate and work ethic, started creating a prototype. I was sure she'd nail it in two weeks. I was soon to realize that my idealistic concept is a slam dunk in theory only.

It's the seemingly easy projects that usually end up stumping you. For example, once we were working on a dry seasoning mix with only four ingredients. It ended up being the hardest prototype we developed for that client. With only four ingredients there was nowhere to hide. We had only four levers to move up or down: move one and the other three changed significantly. My vision for Semisweet Soda was similar: I wanted a clean, short, simple ingredient statement. Most important, I did not want to resort to using high-intensity sweeteners. The idea was to deliver a clean, pure sweet taste—without the bitterness or aftertaste of the nonsugar sweeteners—that was less sweet than regular sodas, and as a result, more refreshing. A secondary benefit of Semisweet Soda would be that it was lower in calories—something that we could use as a nation. But lowering calories was by far second to my bigger industry initiative, which, if successful, could be the model for how to lower the calories in many product categories over the years.

The first few rounds of these prototypes just didn't taste right. Something was missing. Silvina had balanced the other ingredients in the formula—the acidity, the amount of flavor, the amount of color—and we still could not get it to taste good. I started thinking that maybe the reason why we couldn't get this right is that there's some magic carbonated soft drink "golden ratio" out there that we were unaware of. Even if there was, I was intent on breaking the sweetness paradigm. We tinkered and tried and still we couldn't get it to taste right. This prototype became the bane of the project. The other carbonated drinks we were developing for the client—much more complicated in concept—were progressing nicely. Why was making a less-sweet soda so difficult? We tried mouthfeel enhancement, sweetness potentiators, and different sources of acid. With the prototype tasting meeting right around the corner, we put together what we thought was a damn good—though not yet perfect—semisweet beverage and flew out to drink the samples with our client.

Semisweet Soda was killed at that meeting, but not because it wasn't a good idea. Consumers told us they wanted to buy it. The problem (I'm convinced now in retrospect and with sour grapes in my mouth) is that Semisweet Soda was sampled along with four or five other fully sweetened beverages at the meeting. These others set a sweetness norm for the tasters such that Semisweet Soda came up short. Our client had to focus on the products that they could get to market quickly, so Semisweet Soda didn't make the cut even though I pleaded that, with a few more months of development work (Rome wasn't built in a day!), we could have nailed the product. I'm guessing that this very scenario has played out across the beverage industry many times. I'm also guessing that this is why there isn't anything like Semisweet Soda on the market. In this country, at least.

In Japan in mid-2011, Pepsico introduced a product called Pepsi Dry. Upon first sip I felt vindicated. This was exactly the product I had wanted my client to launch. Pepsi Dry measured 5° Brix, which is about half the sweetness of a regular cola, yet it doesn't contain any nonnutritive sweeteners to muck up the purity of the sweetness profile. It's a tiny step in the right direction.

Spend More for Health

The most pressing health epidemics in today's developed world are obesity, hypertension, and diabetes. Each disorder is influenced by what you eat, which is influenced by what you taste and smell. More federal, state, and nonprofit research dollars need to be spent on basic taste and smell research, so that we can understand more about why we make the choices we do.

The fact that baby boomers are now aging may help fund the research done at institutions like Monell. This enormous group of older (read: wealthier) consumers will soon start—if they haven't started already—to experience slight losses of smell (and possibly of taste). They would be the biggest beneficiaries of advancements in cures for smell and taste loss. There's nothing quite like losing one of your senses to spur donations to the cause.

16

Balancing Flavor

In 2004, I got a visit from Pat Galvin, a soft-spoken, talkative man with a marketing background. Pat worked for Levi Strauss & Company, as well as a few major food brands, while he worked for a respected advertising agency. He wanted to discuss a new beverage business.

His new idea was Vignette, a line of sparkling, nonalcoholic beverages made from wine grape juices like chardonnay and pinot noir. Pat appreciates wine, after years of living in the Bay Area, but when his wife got pregnant with their first child, she could no longer share a bottle with him. He realized that there was an untapped niche for nonalcoholic, yet sophisticated, winelike beverages for people who wanted something to drink when wine is not appropriate.

Pat had worked with an independent chef to create the first two varieties, which were good, but he wanted something a bit more sophisticated for his next flavor launch. His goal was to create an adult, wine-like beverage that someone could choose instead of wine: all the complexity and sophistication without the alcohol.

One of the varieties he hired us to develop was zinfandel. At the time Pat and I started working together, I was on a bit of a zinfandel kick, indulging in a glass every night with dinner. This product was the anti–egg salad: something

I could get really excited about developing. Just to make sure you understand the huge distinction between zinfandel wine, which is red, and white zinfandel, which is pink: I'm talking about the red stuff.

I recruited Anne Marie Pruzan, one of Mattson's best food technologists, and her team member Saji Poespowidjojo to create the beverage. First we had to find a supplier of unfermented zinfandel grape juice. There are purveyors of varietal grape juices who sell mainly to winemakers—both commercial and hobbyist—who eventually ferment the juice into wine. We were going to use it in its virgin state.

When it arrived, we quickly learned that unfermented zinfandel grape juice tastes nothing like zinfandel wine. It contains none of the characteristic flavors we were expecting, like cherry, raspberry, black pepper, or earth. Instead, our zinfandel juice tasted like Welch's Concord grape. Not exactly the complex, adult flavor profile Pat wanted for Vignette.

In wine, these complex, characteristic flavors are the result of the fermentation process, which our grape juice wasn't going to go through. We were going to have to create the flavor of fully fermented zinfandel wine from the ground up. Here's how we did it.

First we had to draw the outlines of the Basic Tastes. We started with sweetness. When zinfandel grapes are crushed, the resulting juice is really, really sweet (again, like Welch's). During fermentation, the sugar gets converted to alcohol until there's very little sugar left, and as a result, very little sweetness. We were aiming for a semisweet flavor (the semisweet initiative strikes again!), so we diluted the zinfandel juice with water to achieve the level of sweetness that was appropriate. This became the base of our beverage.

Next we adjusted the sourness. In order to get the flavor right after dilution, we had to add acid to the beverage to get the sourness back in balance. We could have used lemon juice, but we didn't want lemon aroma in our wine country soda, because there's no lemon flavor in zinfandel. Instead, we chose citric acid—the pure acid that comes from citrus fruits and other produce. As its name implies, it lends a sharp sourness akin to citrus, but without the lemony or grapefruity aromas. We'd arrived at citric after we'd conducted a thorough exploration of other sources of acid. We tried tartaric acid, because that's what occurs naturally in grape juice. Not sour or sharp enough. We tried malic acid, the source of sourness in a Granny Smith apple. Too sharp. Citric acid had just the right sour taste profile for our beverage.

After dismissing umami and salt as inappropriate for zinfandel, we moved on to bitter. Good red wines contain tannin, a compound that comes from the skins of the grape, which adds that characteristic red wine mouth-drying effect. Our zinfandel juice had a naturally low level, which we decided was perfect. We'd develop a more bitter, tannic cabernet sauvignon variety in the future for the Tolerant Tasters out there.

Now we were ready to move from taste to aroma. The combination of these two senses, plus texture, would create our zinfandel Vignette's signature flavor. We ordered samples of dozens of natural flavors, including blackberry, raspberry, leather, smoke, green bell pepper, spices, and more. Flavor chemistry allowed us to add blackberry aroma to Vignette, without adding any blackberries or juice.

We continued to tweak the sweetness, sourness, and flavor until we had them in perfect balance. The beverage is then carbonated, which adds a pleasant irritaste that most drinkers read as refreshing. We considered adjusting the color of the drink, but the zinfandel juice on its own, diluted by the carbonated water, was a gorgeous purple-red. It would need nothing else to attract the eye. The result is a complex beverage that tastes of intense—if not quite fermented—zinfandel with a low level of carbonation and a touch of sweetness.

While cooking at home is not the same as formulating a carbonated soda, there are parallels. You still need to craft the Basic Taste outline of the dish. You need to add the aromatic ingredients to fill in the outlines. And you still need to make sure all the flavor elements are in perfect harmony.

There are a few secret flavor ingredients in Vignette Zinfandel Soda that gave us the profile we were looking for. Yet the overall flavor is so subtle that it is almost impossible to figure out what they are. All you know when you savor it is that it tastes just about right.

Recipe Theory

I can be amazingly creative in the kitchen, which is what I do for a living, after all: dream up new food ideas and make them come to life. I have the uncanny ability—which most professional chefs have—to put ingredients together in my head in a way that I know will taste good. However, this does not mean I am always successful in the kitchen.

I also view recipes as rough guidelines. I will look at a recipe and try to figure out what each ingredient is bringing to the party. If the recipe calls for fish sauce, I wonder if soy sauce would work as well to provide umami. If I read a recipe for collard greens I immediately start wondering if the same ingredients and techniques would work with the bitterness of broccoli or kale.

The one thing I simply cannot do is make the same recipe twice. The execution of a recipe is where I fall down. But from experience I've developed a way to ensure that my dinner guests have a good time. In case my main course bombs, I always have a backup, even if it's just the ingredients for pizza or pasta. I also tend to serve a no-fail dessert, like the season's ripest fruit topped with the very best vanilla ice cream I can buy. When guests end the meal with a smile, they tend to overlook failures earlier in the meal. Equally important is good wine that flows like water. Wine is the fuel, lubrication, and current that keeps a dinner party aloft.

I want you to feel equally free to experiment in the kitchen. Each time you cook, make it an educational experience. Don't be afraid to fail. I do it all the time.

Counterpoint Ingredients for Your Pantry

Ingredients for Adding Complexity	Basic Tastes					Aromas	Great for
	Sweet	Sour	Bitter	Salt	Umami		
Seasoned rice vinegar	X	X		X		Pungent, fresh	Salad dressings, soups, sauces, salsas
Soy sauce				X	X	Ferment-y, wine-y	All savory food
White wine	X	X				Citrus, apple, butter, vanilla, oak, floral	Salad dressing, finishing sauces
Fish sauce	X	X		X	X	Fishy, ocean-y, funky	Just a dash livens up all savory dishes
Ketchup	X	X		X	X	Warm spices (clove, allspice, cinnamon), onion, garlic, celery	Smoothing out harsh sour and bitter edges by adding a warm umami roundness
Coffee (soluble)			X			Roasty, toasty, peat-y, earthy, bean-y	Sauces, chilis, confections, baked goods
Cocoa			X			Chocolate-y, fermented, peat-y, earthy, bean-y	Sauces, chilis, confections, baked goods
Celery				X	X	Vegetal, fresh, ocean-y	Adding a light, fresh savory note to sauces, soups, etc.
Parmesan cheese (or other aged cheese)	X	X		X	X	Cheesy, lactic, dairy, nutty, meaty	Adding umami complexity to almost everything savory (at low levels)
Bitter greens			X			Fresh, green, peppery, sulfury	Side dishes, to add contrast to rich foods
Red wine	X	X	X			Cherry, strawberry, woody, smoky, leather, tobacco	Adding a richness and depth to sauces, soups, and dressings

17

Summary: Sensory Truths You Never Suspected

"Ode to Joy," which concludes Beethoven's Symphony No. 9, is one of the most famous pieces of music in the world. If you can't conjure up the melody, or if you just want to add another sensory element to your reading of this book, you can go to www.barbstuckey.com and listen to the most famous bars of it. The entire symphony runs about 65 minutes from start to finish, give or take a few minutes, depending on who is doing the conducting. I'm sure that's too long for some people, but for Leif Inge, a Norwegian composer and sound artist, Beethoven's original masterpiece is about 22 hours and 55 minutes too short. Using digital technology, Inge created his own version. He took the original piece of music and s-t-r-e-t-c-h-e-d it out over 24 hours. His performance art piece, *9 Beet Stretch*, is a full day of Beethoven's Ninth. That's exactly how long the performance lasted when Inge brought his event to San Francisco. It was promoted as revealing "acoustic truths you never suspected."

I spoke with Ryan Junnell, who attended the event. He is a media designer who, at the time he went to hear *9 Beet Stretch*, was very much interested in the concept of slowing things down. He later organized a slow-motion video festival called Slomo Video. Although the San Francisco performance of *9 Beet Stretch* was in 2004, he remembers it vividly. "There was this *huge* sound playing over these *great* speakers. You were surrounded by it," he recalls:

> If you're familiar with even three or four consecutive notes in that piece you would have been astonished to hear those notes stretched out over thirty seconds or a minute. There was a lot of anticipation. You know what's coming next because it's a very familiar piece of music but it may take eight, sixteen, seventeen minutes to get there. So you have this amazing foreplay to your expectation. You're waiting for it, you're waiting for it, you're waiting for it. You can't believe how long it's taking to get there. But then you can savor that slope. It just makes that one note that you love so much, so much more amazing.

He could have been describing tantric sex. The practice of Tantrism is most widely known in Western cultures as the pursuit of a spiritual form of sex that isn't about striving for orgasm. It changes sex from a goal-oriented pursuit into something that's more about the journey. Sex is considered a sacred act "capable of elevating its participants to a more sublime spiritual plane." The root *tan* means "stretch" or "extend" in Sanskrit. Junnell's sensual description of the tantric experience he had at *9 Beet Stretch* got me thinking about how we eat.

Our lives move so fast. We speed through meals that poor people in developing nations would consider a king's feast. We consume more calories between home and the office, while absentmindedly talking on our cell phones, than many people in the world eat in an entire day. We eat while rushing around, and this causes us to eat mindlessly.

Stan Frankenthaler, the James Beard Award–nominated chef from the chapter on umami, now heads up culinary development at Dunkin' Donuts, the quintessential American doughnut shop. His job is to create delicious new menu items that Dunkin' Donuts can sell in their stores. In keeping with the way Americans eat nowadays, Dunkin' has put drive-through windows in many of its stores. Frankenthaler recognizes the limitations of the quick-serve restaurant format. One of his biggest challenges is getting people to think about the food they eat while driving away: "How, when they're running through the drive-through, are you going to get them to stop, for even just a second, and go, 'Wow, that tastes really good'?"

While eating meals mindlessly, we absorb the calories, but we don't absorb the full pleasure they offer. If you ask why we eat this way, we're hard-pressed to explain the real reasons. Who doesn't want to get more pleasure out of the same amount of food? It's the dieter's dream, the hedonist's heaven, a parent's paradise. And all it takes is slowing down.

If you could experience "sensory truths you never suspected," wouldn't you be game? Well, all you have to do is consider each meal a tantric opportunity. If we apply the principles of Tantrism, this would mean that each meal would be stretched out to provide maximum pleasure without pursuit of a goal—for example, finishing—in mind. Why in the world would you want to finish a meal, anyway? It's one of the very few hedonistic, sensual experiences that don't involve dangerous sex, drugs, or alcohol. (Whoops. Dinner should always include wine.) Wouldn't you rather be eating than not eating? If yes, then try to make your meals last a little bit longer. There's research that proves there's something to this.

A Dutch study set out to determine if the speed of consumption of a meal affects how satiated you are at the end of it. The researchers created a menu with four courses. Course one: a lettuce salad with mozzarella, tomato, croutons, and dressing. Course two: macaroni with tomato-meat sauce. Course three: a layered vegetable torta. Course four: a raspberry pudding. Total calories: about 500.

Before they could conduct their survey, they had to control each subject's food consumption before he or she arrived for the experiment. If you were trying to learn about the lunch choices subjects would make during the experiment, it would not be scientifically sound to start with two people who had eaten vastly different breakfasts. Someone who's just breakfasted on Moons Over My Hammy at Denny's (970 calories, including hash browns) will probably respond to lunch differently from someone who's eaten only a banana and a Diet Coke (121 calories if the banana is big). So they instructed all their subjects to drink the same breakfast shake the morning before the experiment.

The researchers brought the diners into the lab and fed them the four-course meal in two vastly different ways. But in both cases the subjects all ate the same amount of calories. The first group of human guinea pigs was fed slowly: their meal stretched out over two hours, with twenty- to thirty-minute breaks between the courses, similar in pacing to a nice restaurant meal. The second group was given thirty minutes to eat the entire four courses; this pace is, sadly, similar in terms of style and timing to the majority of meals we eat in our lifetime. The researchers also measured the levels of appetite-regulating hormones before and after the meal.

After those in the experiment ate their 500-calorie meal, the ones whose courses were separated by breaks rated their desire for food lower than those who ate all of their four courses at once within thirty minutes. Slowing down and eating in a more regimented manner made their post-meal *wanting* lower. This finding is something to celebrate. We are in control of the satisfaction we get from food! All it takes is being present in the moment, paying attention to your food, and eating—enjoying—your food more slowly and more methodically.

Unfortunately, that's not the end of the study. The researchers uncovered a troubling finding in the second half of the test, which says a lot about our contemporary culture. Later, when both test groups were presented with a smorgasbord of waffles, chocolate-coated marshmallows, cakes, peanuts, and salty potato chips, both groups continued to eat despite their previous rating and whether they ate quickly or slowly. After the participants had ingested their 500-calorie meal, whether they'd eaten it all at once or slowly over two hours, they both ate *the exact same amount* after the test. The subjects in the "slow and satisfied group" ignored their lowered desire for food when it was presented to them practically on a golden platter.

Herein lies the rub. Food is everywhere. It is presented to us on golden platters at work, school, cultural events, social events, birthday parties, and more. It is hard to go anywhere in the developed world without encountering delicious food. What you should do is an internal gut check to determine if your wanting for food is physical or situational, and in lieu of those cupcakes, remember your last meal, how it smelled, looked, sounded, felt, and tasted. And take a pass.

When Mindful Eating Doesn't Work

Cornell University's Brian Wansink is dubious about advice to eat mindfully. He has said, "When I talk about mindless eating, some people erroneously say, 'Then the secret to solving mindless eating is to eat mindfully.' No, not if you're ninety-five percent of the population. To eat half of a pea and ask, 'Am I full yet?' may work for some people. And I know calorie counting and preportioning works for some people. But for most Americans, our lives are way too chaotic to accommodate that . . . So for normal people, the solution is not mindful eating. It's to set up our environment, whether at our home or work, so that we mindlessly eat less."

I'll grant him that not everyone has the luxury to spend an hour eating each meal. But just being aware of when you are eating mindlessly is a start! If you can't recall the sensory input you experienced at each of your last three meals, you're probably eating mindlessly. Go ahead: try to remember the taste, smell, appearance, textures, and sounds. If you can't remember any of them in exquisite detail, you are definitely eating mindlessly.

Wansink's research is focused on how to eat less. That's not the point of this book. I'm focused on helping you get more satisfaction out of the food you're eating, assuming that it's the right quantity. But since many people are trying to lose weight, I think it's important to understand how the senses affect how much we eat.

Wansink's research has proved that container size matters. We eat more popcorn from larger buckets, we eat more food from larger plates, we snack more from larger bowls, and we eat more soup from an endless bowl that never empties. The bottom line: If you want to eat less, use smaller containers.

Finding the same sort of chameleon effect at work that I experienced watching my colleague Candice eat a silkworm pupa, he has proved that people eat faster when they eat with other fast eaters, and more slowly when they eat among slow eaters. We mimic the eating behavior of those around us. To eat less, play slow music so that everyone you're eating with will slow his or her pace of eating. Put your fork down between bites. Break your food into courses that you eat one at a time.

Wansink recommends leaving used plates on the table so that you can see what you've eaten (such as baby back ribs or chicken wing bones), and portioning food out onto plates rather than serving it family style. He advocates breaking large bags of food up into smaller portions to eliminate overeating and making healthful food like fruit more easily accessible. The unhealthful stuff he recommends putting in the back of the cabinet, having proved that the little bit of extra effort required to get the candy will persuade some people to forgo it entirely. You cannot underestimate the laziness of the average eater.

Slow Food Versus Slow Eating

I want to make a distinction between the Slow Food Movement—which I endorse—and eating more slowly, for which I am evangelizing in this book. Slow Food "was founded in 1989 to counter the rise of fast food and fast life, the disappearance of local food traditions, and people's dwindling interest in the food they eat, where it comes from, how it tastes and how our food choices affect the rest of the world."

I don't believe that people have a dwindling interest in food. Just the opposite. And I don't think fast food should be demonized. But I do believe in the slow eating of everything you deem worthy (and delicious) enough to eat. To get a taste for slow, complete the Raisin d'Être exercise at the end of this chapter, in which you will stretch out eating a raisin over the course of five minutes.

I did the raisin experiment a few years ago for the first time during a "mindfulness-based stress-reduction class." Before I took this class, my method for reducing stress had been a cold, crisp bottle of pinot blanc. I decided to try a different approach, which some, including myself, might call "woo woo." When I showed up the first day of class, I recognized the look on my classmates' faces. They were just like me—overworked, underslept, unfulfilled—and yet they found themselves, like me, still driving themselves harder. The long and short of this class is that I learned how to meditate to manage stress, which I did, successfully, for about three and a half weeks following the end of the class. But the one thing that stood out for me throughout the interminable meditations and yoga sessions was the first day of class, when our instructor talked us through eating a raisin over the course of five minutes. First we looked at the raisin. Then we felt it. We smelled it. And we ate it so slowly that every chew counted. Not one physical movement was wasted. I vividly remember seeing that raisin's soul. I had never experienced a simple food so completely. This planted in me a desire to approach my food differently, not just the way I managed stress.

About a year later Roger and I were invited to stay at Miraval Resort in Arizona. One of Roger's friends was running the place, and though I'm no fan of the desert, I've never met a luxury hotel that I didn't love. Especially one that promised "life in balance." I had heard about this type of new-agey resort but didn't really know what to expect. Miraval turned out to be sleepover camp for adults. We signed up for activities on a big bulletin board. No kidding: no technology. Roger and I learned how to tame horses with our psyche and attended a couples communication group. The best thing we did all week, though, was attend the "mindful breakfast." We arrived at the Cactus Flower Restaurant along with eight other campers and took our seats. After a trip to the buffet—laden with organic fruits, whole-grain baked goods, and egg white omelets—we all sat down with our plates in front of us. We were then instructed to eat, slowly, being mindful of everything on our plate. Our leader said *Bon appétit* and we got busy being mindful.

Once again, I was a little taken aback. When I gave the food my full, undivided attention, the experience of eating changed from a task to an event. Instead of talking about the day ahead of us, or the activities we were going to do, we paid attention to the food. To this day I remember the feel of the strawberry seeds against my teeth, something I hadn't noticed in years. Both of these mindfulness experiences were instrumental in my ever-evolving relationship with food.

My Tantric Meal

At Mattson we provide breakfast, lunch, and snacks for our owner-employees. Every day there's something different for our "family meal," whether it's Rich's Waffle Wednesday; chilaquiles made by Joaquin; take-out from an Afghani restaurant; Marianne's tomato bread soup; or leftover lamb chops from a project. Regardless, there's always a fresh salad bar, fresh fruit, and a rush to get back to work.

I decided to do an experiment with one of the most mundane of our lunches. On a day when we had the ever-present salad bar and takeout burritos—in other words, nothing out of the ordinary—I invited people to join me for a full thirty-minute mindful lunch. I told them that we'd be eating slowly and deliberately to see what kinds of truths we unearthed.

People took their plates, made their salads, and chose their burrito. Ten of us sat at the big round table in our largest meeting room. I set a timer and we sat in silence for thirty minutes as we stretched our casual lunch out to its breaking point. I never knew thirty minutes could last so long. I had instructed the diners to eat slowly and mindfully, putting their forks (and burritos) down between bites. When we finished, most people still had food left on their plates. If mindful eating did anything, it made us eat less.

When the timer went off, I asked for comments. Julie had experienced a fragrant flashback to her college days when she used to eat burritos frequently. Janine smelled a cucumber and was transported to a luxury spa where she first experienced cucumber water. I noticed that my salad, which I had dressed with raspberry balsamic dressing, looked like a murder scene when I was done, and I was the perpetrator. But mostly I think we were all thankful the experiment was over. We scattered back to our work.

This put me in an awkward position. I wanted to summarize this book with the earth-shattering revelations we had during our tantric meal together. I wanted to write that we all walked away from the table with a newfound lease on life, clearer heads, and satisfied bellies. But this didn't exactly happen. Where was the raisin epiphany I'd had at the stress reduction class years ago? Where were the strawberry seeds between my teeth from Miraval's mindful breakfast? Why were things so different?

I awoke one day the next week with a start. Of course there weren't revelations! We were in an office building with fluorescent lights. We were in a workaday mind-set. The meal was completely unnatural and forced. I found it difficult to sit in silence with these people whom I love. I found it hard to make a burrito last thirty minutes. I found myself uncomfortably trying not to look my friends in the eye lest we laugh and throw everyone off task. Frankly, the whole experience sucked.

I compared my two previous mindfulness food experiences with the one at work. Those previous times, I had been among strangers. One had been in the context of a mindfulness class at an alternative medicine center. The other had been in Arizona at a mind-body spa. I was in the right frame of mind, not worrying about my next conference call.

As a result of this experience, I can't recommend a choreographed mindful meal. If I were writing a book on how to lose weight I might have been okay with the outcome at Mattson. But my colleagues aren't trying to lose weight and that's not at all what this book is about. This book is about using all five of our senses to find the hidden joy in food and in doing so, take the full pleasure out of every bite.

To do this, you don't have to eat in silence. You don't have to limit yourself to "half a pea." You don't have to stretch out your meal. You don't have to be at a spa or take a class. You can practice eating mindfully in the normal course of your life. There are multiple opportunities available to you every day to practice the skill of eating mindfully. The best time or place to enjoy a mindful meal is the one that's *just about right* for you.

Surround yourself with like-minded people you love. Give your food proper respect by sitting down at a table when you eat. Have an open mind and try new foods, occasionally ordering something from a restaurant menu that you normally wouldn't eat. Take in the visual beauty of your food before you dig in. Notice the colors and contours. Appreciate the topography. Let your nose go first. Smell your food before you put it in your mouth. Notice what it's telling you. If it makes noise when you bite into it, consider its signature sound. After you take a bite, let it sit in your mouth a full second before you start chewing. Breathe deeply. Chew and continue to breathe, mouth-smelling the food for as long as you can, noticing the texture of it when it enters your mouth, and as you continue to chew. When you swallow, enjoy that last burst of volatiles that reach your nose and reflect. Notice which of the five Basic Tastes are present. Notice the balance, or lack thereof. If it needs fixing, don't be afraid to fix it. If it's delicious, enjoy. And if you do this mindfully for a few meals, eventually it will become common practice and you'll do it mindlessly. That is the goal.

Taste: The 5-Minute Raisin d'Être

We often take the food we eat for granted. Most of the time we speed through meals, glossing over the details in our race to get calories into our stomach.

The first step in becoming a better taster is becoming a more observant eater. The following exercise will help you slow down, look at an old food in a new way, and think about how you can apply this to the rest of your dining opportunities.

YOU WILL NEED
   A stopwatch or timer
   3 raisins for each person doing the exercise

DIRECTIONS
   1. Set the stopwatch for 5 minutes. Hand out 3 raisins to each taster.
   2. Tell the tasters to eat their raisins in silence over the course of 5 minutes. Tell them to look at each raisin intently before putting it into their mouth. Once they do put a raisin into their mouth, they should keep it there for about 30 seconds before they start chewing. Ask them to try and make the 3 raisins last the full 5 minutes.
   3. Start the stopwatch.

OBSERVE
   • Discuss your experience.
   • What was new to you?
   • What stood out?
   • If you can see something as simple as a raisin in a new way, consider how much more sensory input you could get from a complete meal if you'd slow down and similarly give food your full attention.

18

Fifteen Ways to Get More from Every Bite

1. Chew Well

Consider the amount of force you exert when you flex your chewing muscles. This, in combination with how long it takes you to chew a food before swallowing, may indicate how much flavor you experience. Gentle, prolonged chewing works best. People who use less force (or were less energetic, to use the researcher's terms) and *took longer to chew their food* before swallowing released significantly more aroma molecules from the food. Much of food flavor comes from the volatile aromas. Remember, volatiles are where the action is!

As a result, slower, less goal-oriented eaters get more flavor from the same food than someone who ambitiously attacks it, chews intently, and swallows quickly. Swallowing quickly just makes your meal end sooner.

If you want to release more volatiles, chew slowly and swallow only after you've fully experienced the aroma of the food from the taste on your tongue and mouth-smelling. The sensory pleasure of food is most intense when it is in your mouth. Chew carefully. Breathe. Chew more. Breathe. Then swallow when it's time.

2. Treat Eating the Way You Treat Sex

If you're not eating consciously, you will not pick up on the subtle sensory input that differentiates good food from great. To suck more greatness from your food, make eating the only activity you do during mealtime. Put away the newspaper, turn off the television, and shut down the computer.

Treat eating the same way you treat another sensory activity: sex. If you want to get the most enjoyment from sex, you will pay attention to your partner and revel in the touch, tastes, smells, and sounds you're experiencing. It's unlikely that you'd get as much out of sex while you were watching television, reading a book, or talking on the phone during the act. Multisensory experiences deserve multisensory attention.

Of course, if you're eating alone, you might be tempted to use media to stimulate your brain, but most of what's on television or the Internet will clash with your multisensory meal. If the content of what you're watching is incongruent, you will take less pleasure from the food. Watching the Food Network or *Top Chef* should be just fine, but most of the stuff that's on television will steal your attention from food in general, if not make you lose your appetite altogether.

People eat more food when they listen to the radio during a meal. Unless you're on a quest to gain weight, this is a fairly alarming result of preoccupying your mind while eating. When you listen to the radio or watch television when you eat, the more recently evolved, higher-functioning part of your brain focuses on the media, leaving no resources but the primitive, old part of the brain to process your eating. When you default to the reptilian part of the brain, things like restraint, satiety, and common sense fly out the window.

If you're dining with others, enjoy yourself. But if the conversation gets heated and your heart starts to race, put your fork down. Save the next bite for a less arousing moment. When you're deeply into something, such as the topic at hand, you'll be less able to devote your attention to the food. And if the thing you're deeply into is a new love interest, forget about getting full satisfaction from each bite. Don't waste your money on expensive meals. M.F.K. Fisher wrote that there's nothing more frustrating than cooking a meal for a couple who are newly in love. No matter how deliciously sensual the food, it will pale in comparison with the full-body passion of romantic bliss.

3. Avoid Sensory-Specific Satiety

Sensory adaptation sets in upon the second bite of whatever you're eating. With each additional bite of the same food, you'll get less and less input from it. In other words, if it's delicious, you'll get less deliciousness with every bite. Eventually, you'll tire of it completely, a concept called sensory-specific satiety. It's fairly easy to reverse the effect, though. All you have to do is take a bite of something else. That's why cleansing your palate works. Even if you have a plate of food that's half steak and half broccoli, sensory-specific satiety holds true. If you eat all the steak first, and then all the broccoli, you'll get less enjoyment out of each than if you alternate back and forth: steak, broccoli, steak, broccoli. That's because you're effectively cleansing your palate with each food between bites.

If you're trying to discern minuscule differences between foods or drinks—for example, during wine tasting—it's important to cleanse with a neutral stimulus, such as water or a cracker, which allows your mouth to return to a neutral state so that you can glean the sensory differences between that 2008 Dry Creek Valley Pinot and the 2009.

4. Practice Food Appreciation at Home

The best time and place to teach children about the joy of food is mealtime. Use mealtimes to talk about what's on the table. Teach your kids where it came from, how you cooked it, and what's happening when they put it in their mouths. You can teach your kids the difference between taste and smell using a jelly bean, something they'll enjoy eating as well. When they know the five Basic Tastes, they'll be able to verbalize to you what they taste in their food, what they like, and what they don't. You'll quickly learn whether your children are Tolerant Tasters, Tasters, or HyperTasters. Respect the biology of their taster type, but encourage them to push their taste buds to try new things.

Teach them that tasting is like sports, playing a musical instrument, or math: you have to practice to improve.

If you teach children that they should eat food only at the table, they will begin to respect the meal as an occasion. Even when they snack, make sure they sit down and pay attention to their food. Eating standing up, in the car, or while doing other things like playing computer games or watching television encourages mindless eating, which can have negative health consequences later in life.

Let your kids play with food. Put bowls of sauces, condiments, and jars of spices on the table. Let them find out how things savor. Be their guides as they explore flavor. When they're old enough to accompany you to restaurants, take them out to eat. Encourage them to order from the adult menu, then share with them.

I have heard that some mothers stir into baby food the herbs and spices of their own "flavor principles"—those flavors that epitomize their cuisine. You can also do this to teach them to like the flavor of healthful foods, even before they're fully conscious of them, by stirring a little bit of bitter greens, fishy fish, or whole grains into their food.

5. Be a Star Taster

When you eat, use both hands—not to pick up your food, but to count off the senses one by one (using the Sensory Star in your head) as you consider how each of them is being stimulated. Notice the visual appeal of what you're eating. Make sure you take care to smell the food orthonasally before taking a bite. Notice its texture, texture contrast, and textural changes as you chew. If the food makes noise, consider its signature sound. Next, go through the Taste Star. Which Basic Tastes are present? Which ones are missing but should be present? Are any out of balance? Do you think any are enhancing the others? Could any be suppressing the others?

When you begin to smell through your mouth while chewing, see if you can detect any new and different aromas from what you experienced through nose-smelling. Now that you understand how taste, aroma, and texture work together to create flavor, you can start to get more of it from the food you're eating. Add a little bit of salt. Does it enhance any of the other flavors? Add a bit of sugar. What comes to the forefront and what is masked or muted? Season away! And after every shake, dash, or squeeze, be sure to taste. Taste! Taste! Taste! It's the only way to really learn how flavor interactions work.

6. Don't Accept That Aging Means Less Enjoyment from Food

Simply being aware of how your senses of smell and taste change as you age can help you address any dysfunction you may experience.

Loss of smell will plague most people at some age. The best way to fight it is to employ the other senses. Make sure every plate of food you eat is a rainbow of colors to activate and excite your sense of sight. Enhance your food with a squirt of citrus, which will make your mouth water more, so you'll be able to get more tastes and aromas from the food. Add a crunchy element to your meal to engage your senses of hearing and touch: croutons in salads, cucumbers on sandwiches, and crumbled granola or nuts over just about anything. The more you hear and crunch, the more sensory stimulation you'll experience.

Don't be afraid to use chiles, whether they're dried, smoked, or fresh. They, like citrus, are virtually free of sodium, sugar, and calories. My favorite is ground chipotle chile. A little bit goes a long way toward shifting the tactile system into gear, as well as adding a nice smoky background note. And of course, ramp up the other spices in your food, especially if you can't tolerate heat. Keep shakers of chipotle, cinnamon, smoked paprika, cumin, and basil at the table along with the salt and pepper. If your smell system isn't completely gone, it's possible you might just need to turn up the volume. Be sure to try all of these methods of cranking it up (acid, texture, heat, spices) before you resort to adding more salt or sugar to your food. These two ingredients can result in unintended side effects if you have other health issues like hypertension or diabetes.

7. Know Which Medicine and Medical Procedures Can Affect Taste and Smell

Healthy people generally have healthy smell and taste systems that may decline with age, but this usually happens slowly unless something else is going on. And if something else is going on, it's likely to require medication. Most older people take some if not many medications these days. It helps to know which drugs can reduce sensory stimulation. A wide variety of drug classes can affect smell. This is not a definitive list, but it's a start.

Drugs That Can Cause Smell Disorders

Drug Group with Potential to Cause Smell Disorder(s)	Examples
Calcium channel blockers	Nifedipine, amlodipine, diltiazem
Lipid-lowering	Cholestyramine, clofibrate, statins
Antibiotic and antifungal	Strepromycin, doxycycline, terbinafine
Antithyroid	Carbimazole
Opiate	Codeine, morphine, cocaine
Antidepressant	Amitriptyline, clomipramine, desipramine, doxepin, imipramine, nortriptyline
Sympathomimetic	Dexamphetamine, phenmetrazine
Antiepileptic	Phenytoin
Nasal decongestant	Phenylephrine, Pseudophedrine, oxymetazoline*
Miscellaneous	Smoking, argyria, cadmium fumes, phenothiazines, pesticides, flu vaccine, Betnesol-N
Organic solvents	Formaldehyde, hydrogen cyanide, hydrogen selenide, hydrogen sulfide, n-methylformiminomethyl ester, sulfuric acid, zinc sulfide, pepper and cresol powder, phosphorus oxychloride, sulfur dioxide gas, chromium, lead, mercury, nickel, silver, zinc, cadmium, manganese, cement dust, lime dust, printing powders, silicon dioxide, carbon disulfide, carbon monoxide, chlorine, hydrazine, nitrogen dioxide, ammonia, sulfur dioxide, various fluorides, acetophenone, benzene, chloromethane, acrylates, pentachlorophenol, trichloroethylene
Over-the-counter cold remedies	Zicam Zinc Cold Remedy Nasal Gel, Swabs, and Kids-Size Swabs

*Damage probably requires long-term use.

Source: Christopher H. Hawkes and Richard L. Doty, *The Neurology of Olfaction*, Cambridge University Press, 2009.

The dental surgery for removing wisdom teeth occurs very close to where the chorda tympani taste nerve runs through the jawline. If you've got a painful third molar, this is bad news because your dentist could unwittingly sever your taste nerve. The good news is that damage done by this procedure can be reversed, but it must be done in the first six months following the event. If you are going to have this surgery, keep close tabs on how you experience taste after the procedure. If you notice anything wonky, you'll want to get back in to see the dentist right away. Remember, the clock is ticking on reattaching this nerve.

Many other drugs and medical issues can cause smell and taste loss. The point is to fight the loss. Don't just shrug your shoulders and accept a lifetime of diminishing flavor. Keep tinkering away at the table and in the kitchen until you find something that helps.

8. Protect Your Head

One of the only preventable causes of smell and taste loss is head injuries that result from accidents. In a study of 542 patients who went to the University of Pennsylvania Smell and Taste Center, 20 percent of them had experienced some sort of head trauma. It's important to prevent smell loss because we don't know how to repair it. When it comes to olfaction, a pound of prevention can eliminate the need for a nonexistent cure.

All you need is a case of whiplash or a really bad fall to sever your olfactory nerve slightly, even completely. Wear your seat belt as the best defense. Wear a helmet when you ski, bike, skate, or skateboard. Don't let your kids engage in full-contact sports where they might have their head slammed into the ground. It's important to prevent smell loss because we don't know how to repair it.

Some doctors will claim they can cure your smell loss with medication, which has yet to be proved scientifically. Others tell their patients that smell loss is irreparable, even though the smell system may be repairing itself through constant regeneration.

Ear infections can damage the chorda tympani nerve, the one that runs from the tongue through the inner ear and on to the brain. If you or your children experience ear pain, be sure to treat it early and completely. The same goes for all viral and bacterial infections, including the common flu. The US Centers for Disease Control estimates that one in six people carries the herpes virus, a common cause of loss of taste buds. If you don't know whether or not you have this virus, ask your doctor to test you.

9. Be Adventurous but Patient

Try new foods. It takes five to ten attempts at trying something before you can say you really don't like it. And remember, it's best to be adventurous in a calm, familiar environment. This is especially true for kids. They can handle only so many challenging tasks at a time.

If you want your kids to try something new, make up a punch card with the word *taste* written on it eight times. Or copy and cut out the card I've included here. Each time they taste something, have them punch out one taste. And if they try it eight times, tell them they don't ever have to try it again. Ever. But make sure *you* don't lose patience before the eighth try. Chances are that those eight tastings will have given them some level of appreciation for the food, whether they choose to continue eating it or not.

Being adventurous is also great advice for older adults who are starting to experience taste or smell loss. If the same food you've always eaten starts to taste bland, mix it up. Push yourself to try a new ethnic cuisine like Indian, Thai, or Vietnamese. You'll experience the new flavors innocently with no preconceived notion of what they "used to" taste like. Without a norm, you won't experience the food as lacking tastes or flavors. And if you don't immediately love the new cuisine, be patient. Remember that the same rule applies for adults as for kids. It takes a minimum of five or six tries.

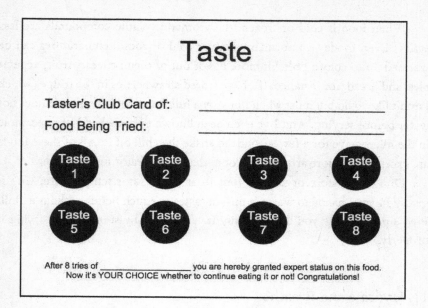

Taste

Taster's Club Card of: _____

Food Being Tried: _____

Taste 1 **Taste 2** **Taste 3** **Taste 4**

Taste 5 **Taste 6** **Taste 7** **Taste 8**

After 8 tries of _____ you are hereby granted expert status on this food.
Now it's YOUR CHOICE whether to continue eating it or not! Congratulations!

10. Taste at the Right Temp

The ever-present microorganisms that take up residence in our food start to die off when food reaches 140°F, so I strongly advise you to heat anything you're cooking to this temperature or higher. Unfortunately, hot food will burn your mouth at about 130°F. I've always got burns in my mouth because I completely lack patience. If you want to get the most out of your food with the least degree of burn, it's best to taste it at about 120°F or lower. Keep in mind that burning your mouth doesn't mean that your ability to taste is diminished, at least not in the long run. Mildly burned buds regenerate in ten to fourteen days.

Don't wait too long to taste your food, though. There's a reason why hot soup tastes better than cold, beyond the fact that you expect soup to be hot. When soup is hot, it releases more volatiles. If you let your soup cool off too much, you will miss many of its wonderful aromas. If you want to experience them as your food is cooling down, put your nose near your plate and breathe deeply.

When food is cold or frozen, the aromatic volatile compounds are less active. Even foods you normally store at cold or room temperatures can be warmed up to coax a little bit more flavor out of them: cheese, fruit, vegetables, and bread, for instance. If I have stored strawberries in the fridge (which I don't like to do but must when I buy by a full flat), I will run them under hot water before serving. And I've even been known to put a bowl of guacamole in the microwave for a few seconds to shake the chill off it. All of these foods are exceedingly more aromatic at room than refrigerator temperature.

Don't ever store or serve a tomato at refrigerator temperature. Cup it gently in your hands to warm it up—it tastes so much better without a chill. Please promise me you'll respect my favorite food by storing and serving it properly.

11. Hydrate and Breathe

Saliva helps liberate the volatile aromas in food. If you notice that you seem to have less saliva than you've had in the past, check your medication. If you can't attribute an unusually dry mouth to a specific medication, see your doctor. Many drugs, even over-the-counter remedies such as Claritin and Sudafed, can dry the mouth and nasal passages. Of course, if you're too stuffed up to taste, they can help you taste more. But if your mouth is too dry, you can end up tasting less.

While water differs in chemical composition from your saliva, it is a fairly good secondary backup. Always have a glass with your meal for those times when you need a little bit more moisture. Of course, you'll use it to cleanse your mouth out between bites, as well, to prevent adaptation.

One long, sustained inhalation works better than a series of short sniffs to suck the volatile aroma compounds far up into your olfactory system where it goes straight to your brain: Do not pass the thalamus! Do not collect $200! The faster and more easily you get the aroma molecules up there, the faster you'll experience the pleasure they give.

Breathing deeply is also important while you're chewing. Our ability to smell through both the nose and mouth makes up for what we lack in olfactory sensitivity.

Breathe in. Breathe out. Circulate the aromas through the system. Have fun with it but don't get too carried away with the laughing or this design feature will reroute liquids up from your mouth and out your nose.

12. Taste First Thing in the Morning

Your taste and smell systems are much more acute early in the morning. They're still virgins. They have not been violated by other tastes and smells for eight hours; therefore you aren't experiencing any taste or smell adaptation, satiety, or cross-contamination. But one sip of orange juice, one bite of banana, one bowl of granola, and your buds have surrendered their virginity. That is, until you go another long stretch without eating.

There are two problems with suggesting that you savor all your food early in the morning: lunch and dinner. There's not much you can do about this, but you can create a virtual virgin palate by not eating, smoking, or drinking anything for two hours prior to your meal. This is the protocol that many research scientists use when they're conducting experiments. If it's good enough for the lab, it's good enough for you.

Here's another tip that can help you get more out of your restaurant experiences. When you dine out, it's common to reach the end of meal and want dessert but realize you shouldn't. Yes, you could share a dessert, but you've been eating for a few hours, so it's likely that it won't taste as good as it would with a virgin palate. Order the dessert to go! The next morning, treat yourself to dessert in lieu of breakfast. Your fresh palate will absorb the full vibrancy of the flavors. Just remember to take it out of the fridge in time to warm it up.

13. Quit Smoking

Smoking can impair your ability to smell. If you're unable to quit—and I don't blame *you*, I blame the nicotine—at least refrain from smoking for at least two hours before your next meal. Your sense of smell is impaired immediately following a cigarette, but recovers quickly. Two hours will do the trick. Unfortunately, most smokers associate smoking with mealtime and look forward to the pre- and post-meal smoke. Try your best to dissociate smoking from mealtime entirely.

If you absolutely must smoke after a meal, wait until the delicious aftertaste has cleared from your mouth. One of the pleasures of a great meal is the warm glow that follows. Why extinguish it early with the taste and smell of ignited tobacco?

14. Use Common Sense with the Scents

To allow your food to take center stage, make the nose-smelling environment as neutral as possible. Avoid strong perfumes, lotions, hair products, after-shave, and air fresheners. If you must, use them with restraint.

If you're over the age of sixty, use them with even greater restraint. If you're prone to smell loss, the odds are that this is the age when you'll start experiencing it. And unless you want to be perceived as Grandma or Grandpa—in an olfactory sense—it's probably a good idea to err on the side of less cologne. If in doubt, ask someone if you've applied the right amount, too much, or too little.

For a few months at work, someone was stocking the Mattson bathrooms with beautiful, froufrou bottles of hand soap. During this time, one of our food technologists brought a sample of a slow cooker meal to my office for my evaluation. I could have sworn I caught a whiff of errant lemongrass in the sample of chicken cacciatore I was about to taste. I lifted my hands to my face and there it was: a noseful of Bangkok emanating from the skin on the front and back of my hands, thanks to the hand soap.

Cooking generates lovely aromas that stimulate the senses. If you've been doing the cooking, it's likely that you won't smell them after a certain amount of time, due to adaptation. Treat yourself by putting down your apron and walking around outside for a few minutes. When you return to the kitchen you will smell your creations anew.

15. If It Doesn't Taste Delicious . . .

In closing, I'd like to quote the immortal restaurant critic Anton Ego, from my favorite animated movie, *Ratatouille*. Ego is a wise man. He gives the best imaginable advice in response to the snarky comment that he's mighty thin for someone who likes food: His reply, "I don't like food, I love it. And if I don't love it, I don't swallow."

A Note About the Notes

Many of the facts and quotes in this book came from personal interviews I conducted over the course of writing this book. These conversations were also supported by reams of published research. For this paperback edition of *Taste*, I have moved the references to my website, solely to save trees. It does not in any way diminish the dozens of incredible scientists whose work made this book possible.

Please visit www.barbstuckey.com.

Illustration and Photography Credits

Illustrations by Russ Cohen: Sensory Star and Taste Star, used throughout; tongue maps, 39, 52; tomato illustration, 58; Sensory Homunculus, 85; Simulation graph, 218.

Photograph by Barb Stuckey: tomato photo, 58.

Photographs by Kristie James: Reinforcements, 20; Plastic cup, 78.

Photograph by Linda Bartoshuk's Lab: Paul Rozin's tongue (with permission from Paul Rozin), 20.

Photograph by Sally Lennon Brown: Bitter face, child, 202.

Acknowledgments

This book would not have happened without the miraculous mojo of my agent, Michael Carlisle of Inkwell Management. He was one of the reasons my editor, the brilliant and patient Leslie Meredith, was willing to take a chance on me as an unproved author. The entire team at Free Press (now Atria) was encouraging, supportive, and professional. A first-time author couldn't have asked for better.

I owe a debt of gratitude to the Monell Chemical Senses Center, in particular Gary Beauchamp, Leslie Stein, Paul Breslin (also of Rutgers), Danielle Reed, Johan Lundström, and Michael Tordoff. Thanks for never making me feel that I was bothering you when even I knew I was.

A huge thank-you is due Linda Bartoshuk, who gave me countless hours of time, always with a smile and passion for the subject, which I hope I have translated to the page. I thank Paul Rozin for having done some of the coolest work in this field, as well as for generously giving of his time.

Thanks to all of the subjects in the book, from people I met through friends of friends to my own friends who were asked to ingest ridiculous (and sometimes illegal) substances for the benefit of the book.

I had two sanctuaries that sustained me during the process of writing. I read most of the papers that informed this book astride one of the spin bikes at the Peninsula Jewish Community Center. And the town of Healdsburg was the light at the end of the tunnel where Scopa's tomato-braised chicken and polenta, the summer sun, and great local wines beckoned.

Thanks to my best friend, Teri Klein, for her unwavering support and professional guidance, as well as Jeff Koppelmaa, a rare combination: my trusted attorney and friend. Thank you Candace Panagabko, whose research supported my effort; Nathanael Johnson, whose help shaped the proposal; and Russ Cohen, whose artwork lightens up the book. Thank you to my readers Chris Patil, Jodie Ostrovsky, mom Joan Stuckey, sorta stepdad Bob Carter, and my sister Zosha Stuckey. Thanks to the entire team at Mattson, especially Candice Lin, Janine Magyar, Doug Berg, Silvina Dejter, and Kristie James, who made me more productive at work so that I could focus my spare time on my writing.

Finally, there are three important men in my life whose support made this book possible, which means they all took more than a few rides on Barb's emotional roller coaster. First, I have to thank Chris Patil, whose friendship is something I will treasure forever, whose editing helped shape the book and whose intellect made sure it stayed non-fiction. Steve Gundrum, the CEO of Mattson, a dear friend and mentor, was kind enough to encourage me to write, but also fatherly enough to encourage me to make some time for myself in the process. And last, I will never be able to thank Roger Bohl Jr. enough for, well, everything from reading and editing to putting up with a partner who had too little time to give him too much of the time I was writing this. He not only holds my hand on the coaster, but talks me down off the tracks more than I'd like to admit. From living with and loving him I have become a better person.

Index

Page numbers in *italics* refer to illustrations.

babies and, 164, 283
blindness and, 114
of chocolate, 140–41
disease and, 170, 171–73
exercises for, 69, 75–81
genetics and, 62
identifying, 8, 68–69
of liquid soap, 202
memory and, 61–62, 68, 69
menstrual cycle and, 168–69
preferences and, 164
pregnancy and, 169–70
scented air and, 66–68
sight and, 107
sight compared with, 60, 107–8
sound compared with, 59
as synthetic sense, 271
taste separated from, 33–34, 49–50
taste's working with, 53–54, 68
two ways for, 55–56, 55
virgin, 151
weight loss and, 9, 301–2
see also aromas
smell loss, 25, 70, 171, 329–32, 336
aging and drugs and, 329–31, 336
weight loss and, 301–2
smell tests, 70
smoking, 335–36
snacks, 304–5, 319, 328
snails, 284
soap, 202, 336
sodium, 159, 160, 174, 185–86, 257, 261, 305–6
reduction of, 188–91, 296–97, 306, 307
sodium chloride, 38, 44, 178, 190, 230
sodium laurel sulphate, 45
soft drinks, 143
semisweet initiative and, 306–9
sugar-free, 216–19
Soltner, André, 5
sorbet, 99
sound, 10, 11, 31, 33, 84, 118–32, 280, 329
blindness and, 114
of chocolate, 141
eating loudly and, 125–27

exercise for, 131–32
of the future, 127–28
hearing problems and, 23, 126–27
isolating, 130–31
marijuana and, 292
misophonia and, 130
music, 121–25, 128–29
noise, 119–20, 127, 128, 130
in restaurants, 119–20, 121–22, 127, 129
sight compared with, 108
smell compared with, 59
taste intensity pegged to, 28–29, 90
of texture, 90
Sound Agency, The, 121
Sound of Music, The (movie), 292
Sound of the Sea, 124
soups:
Campbell's, 189, 191
chicken, 99, 179, 193, 306
French onion, with Gruyère cheese, 259
with glutamates, 253
self-refilling bowl of, 112, 113, 319
temperature of, 99, 333–34
tomato, 88, 259
Sour All Over Tasting exercise, 39, 51–52
sourness, 37, 39, 40, 47, 170, 229–44, 290, 311
aromas associated with, 239
astringency and, 88
as Basic Taste, 31, 33, 35, 37, 39
bitterness and, 200, 206, 212, 214–15
carbonation and, 270
of chocolate, 139
of cola, 216–17
exercises for, 241–44
feeling, 235
ion channels and, 48
location of, 39, 40
mutual suppression and, 182
in orange juice, 45
pairings for, 239
reference food for, 141
saliva and, 229–30
saltiness vs., 48, 237–38
salt paired with, umami, sweet, and, 193

Vignette, 310–12
vinaigrette, 237, 256
vinegar, 37, 172, 206, 235, 238, 239
  balsamic, 177, 235
  rice, 205, 207, 208
  in taste exercise, 39, 51–52
volatiles, 57–58, 72–77, 333–34
  heat and, 74, 75–77, 208, 235, 297, 333–34
  salt and, 181, 183, 192
  in tomatoes, 72–75, 105
vomiting, 30, 46, 64, 84, 283
  pregnancy and, 158, 159, 160

walnuts, 83, 303
Wansink, Brian, 112, 319–20
water, 44, 83–84, 126, 179–80, 229, 281, 311, 334
  acidic, 233
  caffeinated, 141
  dilution solution and, 146, 148–49
  for palate cleansing, 151, 232, 327
  retention of, 296
  salt and sodium and, 161, 185
  sparkling, 151
  spritzing with, 129
  sugar, 224–25
  taste of, 271
water chestnuts, 297
watermelon, 126

Waters, Alice, 71
Waxman, Jonathan, 68–69
wd~50 restaurant, 67, 94
weight gain, 186, 221, 253, 279, 302, 326
weight loss, 9, 171, 221, 301–2
Welch's, 219, 223, 311
Wheeler, Elmer, 118
wine, wine drinking, 3, 8, 10, 16–17, 149, 150–51, 281, 320
  appearance of, 134–35
  brain and, 280, 293–94
  color study of, 107–8
  competitions and, 43, 150–51
  music and, 122–23, 128
  palate cleansing and, 150–51
  pregnancy and, 303, 310
  red, 37, 88–89, 107, 223, 303, *314*
  salt and, 35, 36
  slurping of, 56
  smell and, 54, 69
  taste of, 35
  white, 107, 223
Wolf Children (Malson), 284
Wong, Amanda, 128

Yamaguchi, Shizuko, 253
yogurt, 83, 96

zinfandel, 310–12

About the Author

Barb Stuckey is a professional food developer who leads the marketing, food trend tracking, and consumer research functions at Mattson, North America's largest independent developer of new foods and beverages, where she has worked since 1997. She and her fiancé divide their time between San Francisco and Healdsburg, in California's Sonoma wine country.